Technical Workers in an Advanced Society

Technical Workers in an Advanced Society

The Work, Careers and Politics of French Engineers

Stephen Crawford

Department of Sociology and Anthropology
Bowdoin College, Brunswick, Maine

The right of the
University of Cambridge
to print and sell
all manner of books
was granted by
Henry VIII in 1534.
The University has printed
and published continuously
since 1584.

CAMBRIDGE UNIVERSITY PRESS
Cambridge
New York   New Rochelle   Melbourne   Sydney

EDITIONS DE
LA MAISON DES SCIENCES DE L'HOMME
Paris

CAMBRIDGE UNIVERSITY PRESS
Cambridge, New York, Melbourne, Madrid, Cape Town, Singapore,
São Paulo, Delhi, Dubai, Tokyo

Cambridge University Press
The Edinburgh Building, Cambridge CB2 8RU, UK

With Editions de la Maison des Sciences de l'Homme
54 Boulevard Raspail, 75270 Paris Cedex 06, France

Published in the United States of America by Cambridge University Press, New York

www.cambridge.org
Information on this title: www.cambridge.org/9780521125048

First published 1989
This digitally printed version 2009

A catalogue record for this publication is available from the British Library

Library of Congress Cataloguing in Publication data
Crawford Stephen.
Technical workers in an advanced society: the work, careers, and
politics of French engineers / Stephen Crawford.
        p.   cm.
Bibliography.
Includes index.
ISBN 0-521-35102-2
1. Technicians in industry – France. 2. Engineers – France.
I. Title.
TA158.C73 1989
305'.96'0944–dc19   88-37595   CIP

ISBN 978-0-521-35102-7 Hardback
ISBN 978-0-521-12504-8 Paperback

Contents

Preface

So many individuals have contributed to this study that it is possible to acknowledge here only those whose assistance has proved especially valuable.

My greatest intellectual debts are to my dissertation advisor at Columbia University, Allan Silver, and to my research colleagues there, Peter Whalley and Robert Zussman. Professor Silver provided the initial direction for this project and crucial guidance throughout it. It was a grant to him from the International Division of the Ford Foundation (# 74–0594), that financed the field research in France. Later stages of the research were facilitated by a Fulbright Research Award and a "Summer stipend" from the National Endowment for the Humanities.

At every stage of the research and writing I depended on the kindness of scholars, administrators, secretaries, and others for the help needed in carrying out such a project. In France, the Maison des Sciences de l'Homme (MSH) and Groupe d'etude des méthodes de l'analyse sociologique (CNRS) provided outstanding facilities, as well as administrative assistance and an intellectually stimulating climate. For that I thank Clemens Heller (MSH), François Bourricaud, Raymond Boudon, Terry Shinn, Bernard Lecuyer, Jacqueline Lecuyer, and Pirette Andrés-Puyo. Two other French research groups shared material as well as intellectual resources at later stages of the project, for which I thank Claude Durand, Georges Benguigui and especially Dominique Monjardet at the Groupe de Sociologie du Travail (CNRS), and Marc Maurice, Jean-Jacques Sylvestre, and Arndt Sorge at the Laboratoire d'Economie et de Sociologie du Travail (CNRS). Other French scholars who gave substantial and much appreciated assistance along the way include Jean-Paul Bachy, Luc Boltanski, Mohamed Cherkaoui, François Dupuy, Patrick Fridenson, Guy-Marc Groux, Gérard Grunberg, Henri Lasserre, Dominique Martin, Renaud Sainsaulieu,

Dominique Thierry, and especially André Grelon. Several North American and British scholars provided assistance with the research or with comments on the manuscript, for which thanks go to Robert Blackburn, Val Burris, Loring Danforth, C. R. Day, Duncan Gallie, Ken Prandy, George Ross, Dean Savage, William Schonfeld, Carmen Sirianni, Sandy Stewart, Ezra Suleiman, and Sharon Zhukin. David Burtis was enormously helpful with the designing of the complex programs that produced the time-budget classifications. For the crucial computing facilities and advice provided by the Department of Sociology at the University of Maryland (College Park) I thank Jerold Hage, Reeve Vanneman and the graduate students who worked in the Department's "Data Lab."

It is obvious that this research could not have been done without the cooperation of the two companies involved. Much less obvious are the many ways that local directors, personnel managers, and secretaries went out of their way to accommodate requests for lists, records, documents, interviewing rooms, and tape recorder repairs. I am especially grateful to the engineers who took the time to be interviewed, and to the large number of them who did much more to make my stay pleasant and informative.

The following professional societies, labor unions, trade associations, research teams, government offices, and other organizations also provided timely help at one point or another: FASFID, SFIC, CNIF, CNPF, UIMM, USS (CFDT), UGICT (CGT), UCT, EDF, LEST (CRNS), GST (CNRS), CSO (CNRS), CSO-MACI, ADSSA, CSI (Ecole des Mines), CSE, CEVPFC (FNSP), Entreprise et Personnel, INSEE, Chemical Bank, Centre Culturel Franco-Américain, and the Centre Universitaire Internationale. It is impossible to name the dozens of individuals who went out of their way to make my private life in France more interesting and comfortable, but I want to thank in particular Dominique Monjardet and Antoinette Chauvenet, Pat and Marie-Annick Doherty, and the Gréhant family.

Several research assistants and secretaries have labored on this project. In France, Norma Fuller and Mariane Rupt conducted several interviews each. In the United States, David Buck, Henry Lyons, Susan Lyons, Johnna Major, Janet Morford, and Nicole Polayes assisted with the computing and transcribing. Special thanks go to Ms Morford and Ms Major for their exceptional work as assistants. The secretaries and supervisors in the College typing pool and mail room did far more than required. I am especially grateful to Claire Schmoll for typing the manuscript.

Finally, I want to thank my wife, Liliane Floge. Now an Associate Professor at Bowdoin College, she conducted forty of the interviews in France, provided enormous assistance with the final stages of manuscript preparation, and was remarkably supportive throughout the many years of research and writing.

1

Technical workers in the advanced societies

The decline of the "smokestack" industries and the growing importance of such "science-based" industries as electronics and aeronautics have stimulated much theorizing by sociologists about the consequences for the social structure and stability of advanced societies. Much of the literature has focused on the implications for the social values and collective action of a middle-class occupational group that has expanded rapidly with the rise of the science sector: industrial engineers and technicians. Most of the authors of this literature anticipate a growing resistance by technical workers to their traditional role in industry; some foresee new challenges to the very legitimacy of bureaucratic authority or capitalist society.

Despite such theorizing, only two empirical studies compare the position and values of technical workers in "old" and "new" industry: Zussman's (1985) book on American engineers and Whalley's (1986) volume on British engineers. This study of French technical workers both complements and builds on their important work. It reports on the findings of an investigation into the work, careers, and ideologies of French engineers and managers employed in two industrial settings, a traditional metal-working firm and an advanced telecommunications firm. In the process of assessing and criticizing existing theories, the book also advances its own argument. It shows the significance of a factor largely neglected by all the theories of post-industrial employment, the structuring of careers. It goes on to argue that career structures reflect national institutions as much as any imperatives of industrialism or capitalism, but that such institutions are often peculiar to specific sectors and strata.

This study focuses on France for four reasons. First, recent cross-national investigations of industrial workers (Cole, 1979; Gallie, 1978) and companies (Maurice et al., 1977, 1986) in industrially matched settings reveal large national variations in the organization and experience of work.

Second, the recent books by Whalley and Zussman on British and American engineers provide an opportunity for a study of a third advanced society to identify the distinctively national aspects of the social position of engineers. Third, even a brief examination of engineering in several advanced societies reveals that while French engineers do much the same work as their counterparts elsewhere, the organization of their careers is distinctively French. Only in France are engineering schools part of the elite half of a bifurcated higher education system, and only in France are practicing engineers divided into three fundamentally different categories. Finally, it is in France that some of the most controversial ideas about technical workers in advanced industry have emerged. It is reasonable to suspect that these ideas reflect real developments in France, even if the explanations are wrong or too readily generalized to all advanced or capitalist societies.

Whatever the political significance, it is clear that the technical occupations have grown enormously in recent decades. The United States census of 1980 reports an American population of 1,537,000 engineers. This figure represents an increase of 183 percent since 1950, a much greater increase than the 73 percent for the employed population in general, although slightly less than the 207 percent for that most rapidly growing of the broad occupational groups, "professional, technical, and kindred workers." Within the latter category, engineering is the second largest "occupation" (after public school teaching) and the largest one for males and in industry.

In France, the growth figures are comparable. The French census of 1954 reveals a total of 75,808 ingénieurs.[1] The figures for 1975 and 1982 are 256,290 and 299,000 respectively, about half of which are ingénieurs diplomés, i.e. graduates of engineering colleges. The 1975 figure represents a growth of 238 percent, compared to a growth of 136 percent for nontechnical managers (cadres administratifs supérieurs), 82 percent for lower level administrative personnel (cadres administratifs moyens), 91 percent for office workers, and 14 percent for the labor force in general. The only major occupational category in the French classification that grew more rapidly than engineers, was that of techniciens, up from 193,206 in 1954 to 758,690 in 1975, or 293 percent. Together, ingénieurs and techniciens had gone from representing less than 1.5 percent of the French labor force in 1954, to almost 5 percent (.0466) in 1975. Five percent of the total work force means a considerably higher percentage of the industrial work force. Not surprisingly, the high technology industries have much higher concentrations of technical workers. For example, in 1976 ingénieurs accounted for only 1.4 percent of all steel industry employees but 9.2 percent of all employees in the aeronautical industry (UIMM, 1977: 76, 77, 93, 94).

The second part of this chapter introduces the theoretical debates and empirical issues that inform this study. Chapter 2 describes the research strategy, sites, and samples. Chapter 3 presents the full text of an interview with one telecommunications engineer. Chapter 4 focuses on the knowledge that French engineers use, addressing in particular Daniel Bell's argument about the growing importance of theoretical knowledge. Chapter 5 examines the position of French engineers in the division of labor, giving special attention to theories of deskilling. Chapter 6 looks at the location and attitudes of French engineers with respect to the division of authority, and analyzes the evidence for any decline in the legitimacy of bureaucratic authority. Chapter 7 considers the career structure of French engineering and the attitudes of engineers toward their careers. Chapter 8 examines the collective organization of French engineers, addressing in particular Freidson's theory of professionalization and Mallet's vision of a "new working class." Chapter 9 analyzes the politics and class position of these engineers. A concluding chapter summarizes the findings, and offers an approach to interpreting them that departs in critical ways from both the Durkheimian and Marxist traditions of occupational and class analysis.

Theories

Claims about the emerging dilemmas and reactions of engineers can be conveniently divided into those that emphasize occupation and those that emphasize class. Both approaches view position in the division of labor as the major source of a person's ideology and social participation, but where occupational analysis focuses on the particular tasks people do and the skills involved, class analysis stresses broader categories of authority over property and people. A major issue in the occupational analysis of salaried experts is whether or not they are undergoing "professionalization," that process by which certain occupations become respected and powerful enough to assure their practitioners autonomy in work as well as high salaries and prestige. Class analysis raises several issues, for there are not only Marxist and non-Marxist versions of it, but different variants of each. Most prominent among the Marxist approaches are theories of "proletarianization" and a "new working class." Much less Marxist in its current form is the theory of an emerging "service class" of responsible experts and managers. Finally, there is an important version of class analysis that shares much in common with professionalization theory in its prediction of the rise of a new "professional class" based on knowledge rather than property. This section reviews the key arguments made by the major contributors to this theoretical debate about the place of technical workers in advanced societies.

Professionalization

One reason that American sociologists have long emphasized occupational categories over those of class, is that certain occupations – especially law and medicine – have achieved enormous power and prestige within American society.[2] A recurring claim in the literature on these "professions" is that professional values and norms differ significantly from those of business enterprises and bureaucratic organizations.[3] Professionals are distinguished not simply by the complex and codified character of their knowledge, but also by their exceptional autonomy and authority in the use of that knowledge. Such autonomy and authority stem in part from the public's *trust* in professional claims of commitment to client and public welfare and from the public's confidence in occupational self-regulation.

These alleged differences in professional values and norms are said to generate "value conflict" and "role strain" for professionals who work in bureaucratic settings. Kornhauser (1965), identifies several such conflicts among scientists in industry, the problem being that industrial research is shaped and scheduled by company executives who are not scientists and whose goals are profit, not the expansion of human knowledge. The normal reward for faithfully serving the bureaucracy is vertical mobility within the organization, but local mobility into management is of little interest to professionals, for whom success involves broader recognition within their occupation.

Kornhauser's study, however, was limited to scientists and engineers working in research laboratories. The question is whether or not his model of conflict between professional and organizational values is applicable to the majority of industrial engineers. There is considerable evidence that it has not been in the past. On the basis of a large sample survey, Goldner and Ritti (1967: 491) report that "Engineers generally enter industry with non-professional goals . . . are oriented towards entrance into positions of power and participation in the organization rather than simply practicing their original specialties . . . and strongly identify with the organization and its goals."

This is not surprising when one considers the long history of career mobility into management for American engineers. In his study of the career patterns of engineering graduates from 1884 to 1924, William Wichenden (Noble, 1979: 41) found that within fifteen years of graduating from college, about two-thirds of these engineers had become managers. More importantly, such mobility into management represented *professional* success in the eyes of engineers. Consequently, as David Noble (1979: 42) observes "the potential conflict between professional integrity and subordinate corporate employment did not arise simply because their unique notion of professionalism was one which neatly embraced corporate position as the mark of status within the profession." On the basis of his study of engineers in several European countries, Torstendahl (1982: 262) similarly argues that the industrial engineers emerge as both "professional men" and "new bureaucrats," and that professionalization and bureaucratization "did not in any obvious way prevent each other's development."[4]

Research on more contemporary engineers supports a view of them as predominantly "local" in orientation rather than "cosmopolitan." Perucci and Gerstl (1969: 179) emphasize the lack of a professional community among engineers and the "relative absence of major conflict between organizational and professional norms concerning autonomy." In short, by the early 1970s, sociologists (Carter, 1977; Ritti, 1968, 1971; Rothstein, 1969; Wilensky, 1964) were largely in agreement that analyzing engineers as professionals shed little light on their particular dilemmas and attitudes.

Recently, however, two influential theorists of post-industrial society have rekindled the role strain argument in the distinctive context of the expanding, science-based industries. Many of the writers about post-industrial society (Lane, 1966; Etzioni, 1968; Galbraith, 1968; Bennis and Slater, 1968; Touraine, 1971) point to the growing importance of "knowledge-based" work, but Daniel Bell (1972, 1973) and Eliot Freidson (1973a, 1973b) stand out for their emphasis upon the changing role and social significance of knowledge in production. For Freidson, moreover, this strain is rooted in the professionalization of knowledge-based workers.

Freidson observes that in the emerging post-industrial society, a growing proportion of production workers are "knowledge-based" workers whose increasingly abstract skills require long periods of formal training in specialized schools. He claims that "such higher vocational education does not merely insert knowledge into people's heads, but also builds . . . occupational identities and commitments" (1973a: 52). The result is the development of "occupational solidarity," among the workers practicing the same specialized skill. In Freidson's words (1973a: 52):

Their skill is not merely abstractly there as a potential, but it is institutionalized as a stable discipline or occupation. Such trained workers do not constitute a class of labor which can be treated as mere hands, to perform whatever tasks management may invent for them and then train them for. Rather, they are a kind of labor with pre-existent skills for which management may have a need but which management must take more or less as given. Their tasks are institutionalized occupationally, and thus resist simplification, fragmentation, mechanization, or some other mode of managerial rationalization of labor.

The capacity to resist management pressure toward rationalization and integration is reinforced by whatever degree of monopolistic labor market control the occupation can gain through control over the recruitment, training, and licensing of its practitioners (Freidson, 1973b: 27). It is also reinforced by the fact that the new "knowledge-based work, the work of middle class experts, professionals and technicians, is *by its very nature* not amenable to the mechanization and rationalization which industrial production and commerce have undergone in the past century" (1973a: 55). For these various reasons, management is unable to fully control the labor process. Rather, knowledge-based workers and their occupational associations will increasingly define tasks, determine who is qualified to perform them, and control and evaluate performance. In view of the high concentrations of knowledge-based workers in the science-based industries, it follows that such challenges to bureaucratic authority will be most pronounced in the high technology sector. This study tests such expectations by comparing the occupational identification and the attitudes towards bureaucratic authority of French engineers in low and high technology settings.

Proletarianization

Classes group together persons engaged in a wide variety of occupations on the basis of shared location in the social hierarchy. In Marxist analysis, this hierarchy is a structure of economic power characterized by dominant and subordinate positions. Technically speaking, proletarianization refers to downward movement by an individual or group in this hierarchy of classes. Thus, the movement of peasants and self-employed artisans into jobs as employees of someone else constitutes proletarianization. However, most Marxists recognize that among dependent employees there are great variations in both market power and work autonomy, and that these in turn are linked to skill level. In this context, proletarianization has come to refer to declines in the skill and autonomy of work that result from management efforts to better control the labor process. It is this second meaning of deskilling that Marxists have in mind when they apply the term proletarianization to engineers.

The most influential recent statement of the proletarianization argument is Harry Braverman's *Labor and Monopoly Capital*, published in 1974. Although Braverman's book focuses on the historical deskilling of craft and clerical workers, its chapter on the current "middle layers of employment" includes a discussion of engineers and technicians. Braverman (1974: 243) argues that "having become a mass occupation engineering has begun to exhibit, even if faintly, some of the characteristics of other mass employments: rationalization and division of labor, simplification of duties, application of mechanization, a downward drift in relative pay, some unemployment, and some unionization."

This is about all Braverman has to say about engineers, but several other sociologists have applied the deskilling thesis to professionals and managers. Of particular interest are recent articles by Derber (1982; 1983), Meiksins (1982), Larson (1980), and Bauer and Cohen (1980; 1982). Derber (1982: 8) claims that "like other workers," professional employees "have become detail workers, unable to choose their own projects or tasks and forced to work at the rhythms and procedures institutionalized in the job descriptions and standard operating procedures of the organization." Although not yet subjected to the intensive rationalization and control of "technical proletarianization," they suffer from "ideological proletarianization." Engineers, for example, "are deprived of their right to select and formulate their own research objectives," and experience a "loss of control over the organizational uses and application of their technical investigations" (1983: 319, 321).[5]

Proletarianization theorists who make more straightforward claims about the fragmenting, deskilling and routinization of engineering work include Carchedi (1977) in England, and Bauer and Cohen (1980; 1982) in France. Groux (1985: 31) cites Bauer and Cohen in attributing recent patterns of *cadre* unionism to "an increasingly taylorian organization of intellectual labor, which implies a modification and fragmentation of the tasks of *encadrement*, a reduction in their contents, and a degradation of the conditions of work of *cadres*." (The word *cadres* is best translated into English as managerial and professional employees.)[6] Bauer and Cohen, however, also emphasize the organization's appropriation of technical knowledge and the "crystallization" of such know-how in manuals of procedure. In their words, "it takes only a little experience inside large industrial firms to discover the multitude of bibles which specify very precisely, for all areas of work, exactly how to formulate and resolve" the problems met by engineers (1982: 459).

It is not only academic sociologists who make such proletarianization arguments. The French Communist Party (PCF) offers a very similar analysis of the fate of "middle strata" under "state monopoly capitalism" (Ross, 1978). But while the proletarianization theorists neither distinguish between low and high technology industries, nor claim that objective proletarianization is producing working-class politics, the PCF does both. According to it, deskilling and wage reductions don't make professional employees members of the proletariat as long as they remain engaged in unproductive labor, but do make them more disposed towards alliances with the working class. Moreover, for certain white-collar workers, routinization of their work is associated with the emergence of the "collective laborer" in the technically advanced sectors, by virtue of which many engineers and technicians "become productive workers and hence bonafide members of the working class" (Ross, 1978: 168). In an effort to test these theories of proletarianization, the following chapters examine and interpret developments in the work, pay, unionization, and politics of French engineers.

New working class theory

The PCF analysis of the "middle strata" was roundly criticized by other French Marxists. Nicos Poulantzas (1978) argued that the PCF's impatience to recruit electoral allies was blinding it to the inherently middle-class position of engineers and middle managers, and leading to social democratic compromises in the party's traditional revolutionary goals. More influential, however, was a group of critics that made almost the opposite argument. These "neo-Marxists" questioned both the progressiveness and revolutionary capacity of the modern proletariat, and the applicability of the proletarianization thesis to knowledge-based workers in advanced industries.[7] Nevertheless they saw in these knowledge-based workers the vanguard of a radicalized "new working class."

France's post-war economic "miracle," including the surge in workers' real incomes and the rapid growth of non-manual occupations, generated serious doubts on the French left about the more orthodox Marxist theories of proletarianization and pauperization. Liberal sociology's notions of working-class "embourgeoisement" and middle-class professionalization offered unappealing alternatives, as did the "critical theorists'" discouraging view of technical workers as increasingly effective technocratic legitimators of the existing social system.[8] In this economic and intellectual context, a handful of innovative French leftists – Mallet, Gorz, Touraine, etc. – developed in the 1960s a new analysis of the technical stratum, one dispensing with such shaky notions as proletarianization and "false consciousness," but giving a radical turn to liberal theories (Blauner, 1964) of worker integration in advanced industries.[9]

The boldest and most influential statement of "the new working class" thesis is the late Serge Mallet's book of that name (1975).[10] Mallet thought that workers in the high technology sector are not only more skilled, secure, and highly paid than workers in traditional mass-production industries, but that they also enjoy greater autonomy and professional challenge in their work and more involvement in small work groups that cut across traditional hierarchical differences of rank. Yet, he did not believe that such "objective integration" leads to subjective integration in a capitalist firm. On the contrary, it accentuates the worker's experience of certain "contradictions" inherent in capitalism, especially in the case of technical workers.

The most important contradiction is that between the job and the organization. The job of a technician entails autonomy and involvement; it invites and rewards creative participation in production. The organization, on the other hand, is bureaucratic and hierarchical, and technicians are permitted little if any voice in its decision-making processes. Such powerlessness in a work situation that is otherwise involving rather than alienating creates an inconsistency that gives rise to greater frustration and militancy than found among traditional workers.

A second contradiction concerns the strategic position of technicians in science-based industries, and recalls Veblen's (1965: 69) analysis of engineers as the "indispensable General Staff of the industrial system." By virtue of its technical knowledge and central role in the production process, the technical staff understands the techniques of manufacturing well enough to manage without guidance from the official managers. And by virtue of its critical place in the actual operation of factories, the technical staff is in a strong position to challenge the status quo. Aware of these technical and political capabilities, the technical stratum thus experiences capitalist management as all the more gratuitous and intolerable.

Third, there is a value conflict here that calls into question the legitimacy of capitalist management. Again new working class theory resembles the much older analysis by Veblen, who saw the relationship between engineers and capitalist managers as inherently conflictual. Veblen (1965: 74) argued that sooner or later the engineers would revolt and take command, not only because they have the capability, but also because "by training, and perhaps also by native bent, the technologists find it easy and convincing to size up men and things in terms of tangible performance, without commercial afterthought," and thus "are beginning to understand that commercial expediency has nothing better to contribute to the engineer's work than so much lag, leak, and friction." Similarly, new working class theorists (Gorz, 1967: 103–4) maintain that the subordination of research, development, and knowledge diffusion to the logic of capitalist profit-making violates the "higher rationality" and service orientation implicit in technical work and emphasized during education in the sciences. In short, by virtue of their culture of technical work, engineers are anti-capitalist.

Mallet, Touraine and Gorz mention other contradictions, and disagree about some of them. They agree, however, that the essential problem is organizational alienation rather than economic exploitation, and that as a result, the new working class is less concerned than the "old" about wages, but much more concerned about control. *Autogestion* – workers' self-management – is the characteristic demand of the technical and skilled workers in the high technology sector. The interest in democratic control also manifests itself in new forms of anarcho-syndicalist and plant-specific unionism (*syndicalisme d'entreprise*). In Mallet's and Gorz's versions, the class character of this movement means challenges to capitalist rule in the political arena as well as in the workplace. *Autogestion* within the plant is but the first step in an unfolding social revolution, one that builds socialism from the grass roots up rather than through the seizure of state power.

New working class theory was widely discussed and debated in France during the 1960s and early 1970s, especially after the student riots and general strike of May–June, 1968. Gorz himself changed his view. In a passage that echoes C. Wright Mills's pessimistic conclusions to *White Collar*, Gorz writes: "More often than not the rebellion of intellectual workers is profoundly ambiguous: they rebel not *as* proletarians but *against* being *treated* as proletarians." Aimed at the "reinstatement of the privileges they once enjoyed as members of the professional 'middle class'," such struggles "are, in fact, anti-monopolist rather than anti-capitalist in character . . . and as likely to become fascist as reformist" (1976: 178–182).

Non-Marxists were equally skeptical. The British social theorist, Anthony Giddens (1973: 196), criticized some of the basic premises, especially the claim that the groups referred to in any way constituted an actual class. On the American side of the Atlantic, Daniel Bell (1973: 154) ridiculed new working class theory as "simply a radical conceit, and little more."[11] Nevertheless, this theory remains influential and valuable, especially in France. Thus, the chapters on authority, unionization, and politics analyze this study's evidence for and against it.

Professional class and service class

Like Freidson and others, Daniel Bell (1973: 20) sees "post-industrial society" as distinctively knowledge-based:

Industrial society is the coordination of machines and men for the production of goods. Post-industrial society is organized around knowledge, for the purpose of social control and directing of innovation and change; and this in turn gives rise to new social relationships and new structures which have to be managed politically.

What distinguishes Bell's argument, however, is his (1973: 20) emphasis on the increasing significance of theoretical knowledge, "the primacy of theory over empiricism and the codification of knowledge into abstract systems of symbols." Bell argues that the post-industrial industries of computers, electronics, optics, and polymers are "science-based" industries in a way that the steel, automobile, telegraph, and electricity industries are not. The latter "were developed largely by talented tinkerers who worked independently of the fundamental work in science" (1973: 116). By contrast, the science-based industries "are primarily dependent on theoretical work prior to production," and thus represent a changed and newly intimate relationship between science and technology (1973: 25).

As theoretical knowledge becomes more central, so also do its appliers, creators and diffusers: "if the dominant figures of the past hundred years have been the entrepreneur, the businessman, and the industrial executive, the 'new men' are the scientists, the mathematicians, the economists, and the engineers of the new intellectual technology" (1973: 344). Bell (1973: 374, 362) concludes that "the major class of the emerging new society is primarily a professional class, based on knowledge rather than property," and its norms, "the norms of professionalism are a departure from the hitherto prevailing norms of economic self-interest which have guided business civilization."

Unlike Freidson, however, Bell does not anticipate a significant challenge to bureaucratic industrial authority from the technical intelligentsia.

There are several reasons for this. The most important is that the "societal structure" of post-industrial society fragments the professional class and even its constituent "estates" and occupations in ways that inhibit occupational identities and solidarities. The professional class is divided into four estates: the scientific, the technological (applied skills: engineering, economics, medicine), the administrative, and the cultural (religious and artistic), and although "bound by a common ethos, there is no intrinsic interest that binds one to the other, except for a common defense of the idea of learning; in fact there are large disjunctions between them" (1973: 376). Moreover, the members or any one estate, or even of any one occupation, tend to work in several significantly different and increasingly competing *situses*, such as economic enterprises, government agencies, universities and research institutes, social complexes, and the military, and this of course inhibits estate or occupational consciousness. As workers whose expertise is in the application of codified knowledge to social or economic purposes, technologists in particular "are constrained by the policies of the different *situses* they are obedient to" (1973: 376).

In using the word "constrained," Bell does not mean to imply severe tension between professional and organizational norms. In Bell's view, management also is becoming more professionalized; the managers belong to the administrative estate of the professional class. And contrary to the large disjunctions between certain estates, the administrative and technological estates share a common commitment to planning, efficiency, and functional rationality. Moreover, Bell (1973: 324) believes that "in the coming decades the 'traditional' bureaucratic form will have given way to organizational modes more adaptive to the needs [of technical and research personnel] for initiative, free time, joint consultation, and the like," thus further reducing the potential for conflict between professional employees and management. Finally, the growing centrality of theoretical knowledge is altering the relative power of business, government, and science in such a way that "the *situses* rather than the [occupational or socioeconomic] statuses would be the major political interest units of the society" (1973: 377). In short, class conflict is yielding to the politics of *situs*. Chapter 4 of this study examines some evidence bearing on the kinds of knowledge used by French engineers in traditional and high technology industries. Chapter 8 considers the evidence for a politics of "*situs*".

Related to Bell's theory of an emerging "professional class" are "new class" arguments. These share a common emphasis on the rise of a professional, technical, and cultural intelligentsia whose interests and ideology reflect their possession of educational credentials rather than productive property. Yet, it is important to distinguish two variants of them. Neoconservatives – i.e. Moynihan, Kristol, and Wildavsky (Bruce-Briggs, 1981) – criticize what they view as the "adversary culture" radicalism of a self-seeking counter-elite. By contrast Gouldner (1979) sees the "new class" as a potentially progressive force which, thanks to its culture of "critical discourse," is capable of acting in behalf of societal as well as its own class interests. Although secondary issues in this study, these new class themes are addressed briefly in the analysis of political ideology in Chapter 8.

More central to this book's main argument is the last concept of class raised here, the largely neglected concept of a "service class." Building on ideas developed by Renner several decades ago, John Goldthorpe (1982: 168) argues for treating all employees "to whom authority is delegated or to whom responsibility for specialist functions is assigned" as members of a single class of "trusted workers." This class is distinguished from the working class by the moral content and long-term nature of its relationship to employers and by the promise of career advances it receives "in return for the acceptance of an obligation to discharge trust 'faithfully'" (1982: 169). Such trust is required of workers whose jobs necessarily entail considerable autonomy and discretion. According to Goldthorpe, engineers belong to this "service class," but technicians belong to the "intermediate strata" located between the service and working classes.

In his book on British engineers, Peter Whalley (1986) makes a strong case for analyzing engineers as "trusted workers," but gives less attention to the more controversial issue of a "service class." On the surface, the concept of the service class appears especially applicable to France, where engineers, financial administrators, personnel managers, sales executives, and general managers share the important status of *cadres*. The chapters on authority, careers, unions, and politics explore the utility of viewing French engineers as members of a service class.

Empirical studies

Sociologists and historians have produced a large body of empirical research on engineers, *cadres*, and technicians, some of which addresses the theoretical issues of concern here. Several of the best works have appeared since this study began, but have enriched the analysis and interpretation of its findings.

Historians and historical sociologists have shed much light on the collective efforts of engineers to improve their status and market power. The works of Calvert (1967), Calhoun (1960), Layton (1971) and Noble (1977) describe the largely unsuccessful struggles of American mechanical, electrical and other engineers to develop codes of ethics, unify, and form an association powerful enough to effectively represent their interests to employers and the state. In *The Rise of Professionalism* (1977), Magali Larson shows how the subordination of the market for engineers to that for products produced by large manufacturing companies made it virtually impossible for engineers to gain the kind of market control achieved by physicians.

Histories of French engineers show only slightly greater success. The works of Boltanski (1982), Grelon (1986), Mouriaux (1984), Shinn (1980a), Thépot (1985), and Weiss (1982), show the preoccupation of French industrial engineers with achieving the kinds of status and privileges long enjoyed by the exclusive corps of state engineers, the gradual emergence of engineering associations representing salaried "civil" (i.e. non-state) engineers, and the role of the engineering schools and their well-organized alumni societies in efforts to restrict the practice of engineering to certified engineers. The latter effort was partially successful, for French engineers did obtain legal protection of the title *ingénieur diplomé* in 1934 and considerable influence on the state body that regulates engineering education. Yet, by mid-century the most powerful associations representing French engineers were unions of *cadres* that represented salaried administrative as well as technical workers.

A great many French sociologists examine the contemporary status and identity of engineers and *cadres*, and a few of these authors (Benguigui and Monjardet, 1968; Benguigui et al., 1978; M. Durand, 1972; C. and M. Durand, 1971; Maurice, 1969) address questions relevant to the professionalization debate. However, none of them focuses on the issues raised by Freidson, and none of them compares engineers in old and new industry. Whalley (1986) and Zussman (1985) do make such comparisons in the British and American contexts, and they find little support for Freidson's expectations. Whether their findings apply to France, however, remains to be seen, for the history of French engineers demonstrates their capacity to develop quite distinctive "collective mobility projects."

Of the empirical works on the deskilling of technical labor, most of the American studies are critical of the Braverman thesis. The major exception (Kraft, 1977) focuses on an unrepresentative group, computer programmers. Zussman emphasizes the authority that many engineers continue to exercise on the job, even in high technology settings. He acknowledges that this is less the case for the growing numbers of research and development engineers, but points out that the latter use their technical skills more than other engineers, enjoy considerable autonomy in doing so, and are not suffering declining wage differentials or joining unions. Whalley emphasizes the disinterest of employers in deskilling "trusted workers," and the resulting tendency to assign any deskilled technical tasks to technicians. Meiksins, who earlier (Derber, 1982) stressed the rationalization and control of technical work, now finds that 91 percent of the engineers in his recent sample survey (Meiksins and Watson, 1987) "are usually or always able to decide how to go about their jobs" (p. 9) and "certainly have not been transformed into detail workers" (p. 14). Smith (1987: ch. 4) emphasizes the skill and autonomy of British technical workers in the aerospace industry.

The French literature, however, paints quite a different picture. From the case studies of Bauer and Cohen (1979, 1980, 1982) mentioned above, to the statistical analyses of Baudelot et al. (1974), to the historical investigations of Dejonghe (1985), the accent is on the decline of *Monsieur l'ingénieur*. Where he was once the employer's right-hand man, a technical innovator and a manager of production, an officer at work and a respected gentleman in town, he is now often viewed as a skilled worker. The theme is echoed in a large number of books on *cadres* (e.g. Cheverny, 1967; Dubois, 1971; Lasserre, 1984; Quin, 1976). Yet, many French authors would agree with Jean Saglio (1984: 349) that "an engineering diploma is the best way of getting to the top of industrial firms for those who are not lucky enough to be the proprietors."

A few empirical studies address the themes of a "professional class," a "service class" or simply a "new class," but except for the work of Zussman and Whalley, none to my knowledge are based on investigations of engineers. Zussman examines the knowledge engineers use, and finds little support for Bell's argument about the growing centrality of theoretical knowledge. Whalley makes good use of the concept of a "service class," but Zussman is equally convincing in his treatment of engineers as members of a broader "working middle class" that is structured by its members' positions in the community as well as at work. Boltanski (1982) views French engineers as the central movers in the formation of *cadres* as a middle-class "third force" between capital and labor.

Several empirical studies attempt to test new working class theory; their findings are contradictory and inconclusive. Barrier (1968) interprets her data as damaging to Mallet's argument, and Sainsaulieu (1972) finds little in the way of militant attitudes among white-collar technicians, but a number of careful studies by the Durands (C. Durand, 1968; 1971; M. Durand, 1972; Durand and Durand, 1971) provide some support for the new working class thesis. Low-Beer (1978) finds that Italian technicians do seek more control at work, but attributes this to their social and political origins rather than their work situations.

Many, but not all, of the French studies (Maurice and Cornu, 1970; Willener et al., 1969; C. Durand, 1971; Dulong, 1971) conducted after the strikes of May–June 1968 cast serious doubts on new working-class images of technicians and engineers. Gallie (1978) finds a good deal more evidence for radical attitudes among his sample of refinery workers, but contrary to the new working class explanation, he attributes this pattern to factors that are specific to France, such as managerial paternalism and the industrial relations system. Jean-Daniel Reynaud (1972) also questions the causal connections between technology, jobs, and attitudes, and emphasizes the influence of such intervening variables as organizational structures and personnel policies. He argues that technicians identify more closely with the *cadres* above them than with manual workers, and that their anger reflects blocked mobility aspirations, not class consciousness.

Only three major studies compare engineers in old and high technology settings. The two of these that do so most systematically are Zussman's (1985) and Whalley's (1986) books on American and British engineers. Neither finds support for Mallet's expectations; both emphasize the similarities in the attitudes of traditional and high technology engineers. However, Grunberg and Mouriaux's (1979) careful analysis of a national survey sample of French *cadres* reveals significant differences in political attitudes between old industry and science sector engineers. So does this study. Such cross-national variation points to the need for research that looks beyond low and high technology.

Although brief and selective, this review of the empirical literature shows the broad disagreement among researchers who have conducted empirical studies relevant to the theories under consideration.[12] The two major investigations of engineers in old and new industry offer important criticisms of most of these theories, but criticisms based on studies of American and British engineers. Cross-national studies (Cole, 1979; Gallie, 1978; Maurice et al., 1986) of matched industrial sites point increasingly to the variations among advanced societies. It is the nature and sources of these differences that are so problematical, and to which we turn now.

Research issues: labor markets and nations

Both liberal and neo-Marxist theories of post-industrial society assume that production technology has determinate effects on the division of labor, and treat the division of labor as the major source of group identification and political ideology. Such analyses ignore other important sources of work experience, economic interests, ideology, and the division of labor itself.

An important tradition in sociology criticizes division of labor determinism. One alternative is to look for the sources of social identities in the nature of primary groups, community situations, and political institutions (Janowitz, 1978: 148). Another is to consider the effects of workers' "prior orientations" (Goldthorpe et al., 1969) on their experience of work.[13] Low-Beer (1978: 230), for example, criticizes new working class theorists for assuming that the sources of technicians' militancy are exclusively internal to work organizations, claiming that his findings "show strong causal relationships only in the opposite direction: from attitudes acquired outside the work situation to militancy in the work situation." Even some Marxists argue that managers and engineers occupy such "contradictory" (Wright, 1979; Carchedi, 1977) or "structurally ambiguous" (Crompton & Gubbay,

1977) positions between the proletariat and the bourgeoisie that their class situation is "underdetermined" by their position in the relations of production.

Yet, non-work factors are not the only possible way to account for variations in the industrial and social attitudes of individuals occupying the same position in the social and technical division of labor. Variations in market situation may be equally important. For technical and many other salaried workers market situation refers to more than immediate earnings capacity and unemployment risks, it refers to career prospects. An important tradition (Hall, 1948; Wilensky, 1960; Spilerman, 1977) in the sociology of occupations shows the significance of careers for social integration, but tends to ignore career variations among persons in the same occupation. Yet, the career prospects of two French engineers doing exactly the same job may differ considerably according to their age, sex, and education. A central question addressed in this book concerns the consequences of differences in career prospects for the ideologies of French engineers.

If individual careers are important, so also are labor markets and the forces that shape them. It is probably true that production technologies largely determine the *tasks* that must be accomplished in order for industrial enterprises to manufacture given products "efficiently." It is less clear, however, that either the logic of industrialism or that of capitalism determines how tasks get combined into *jobs*. In an important study of Japanese and American auto-workers, Cole (1979) shows that job specification is much greater in the United States than Japan, and attributes the difference to nationally distinctive historical developments in labor–management relations.

If job structures within the same industry may vary by country, it seems all the more likely that neither the stages of industrialism nor those of capitalism determine how jobs are linked into career lines. More than tasks or jobs, career lines imply judgments about persons and their personal qualities. The age, sex, seniority, and educational credentials of job occupants are often better predictors of their careers than is their first or current job. Career lines thus imply social definitions of worthiness for promotion and social systems for producing and certifying those deemed worthy. To the degree that jobs involve trust as well as technical competence, there would seem to be all the more room for national variations in the social production and recruitment of trusted workers. A growing literature (Osterman, 1984; Rubery, 1987) on internal labor markets points to variations in hiring, promotion, and pay policies across firms, industries, and nations, and goes on to criticize both neoclassical and Marxist explanations of structures of employment as overly deterministic.

For the purposes of this study, the most important work in this literature is *The Social Foundations of Industrial Power: A Comparison of France and*

Germany, by Marc Maurice, François Sellier, and Jean-Jacques Sylvestre (1986). Building on recent critiques of radical labor process and labor market segmentation theories as well as critiques of neoclassical economics, Maurice et al. (1986), Rubery (1987), and others insist on viewing the firm as an autonomous actor rather than simply a passive agent of either market forces or collective capital. Yet, Maurice et al. (1986: 224) also reject the micro-based contingency theorists of organizational structure for tending "to reify both the concept of organization and the concept of technology by neglecting the extent to which both are social constructs." Rather, they attempt to develop a synthesis of micro and macro approaches to labor market structure that views employers as relatively independent in structuring employment, but only within the framework of the institutions of the larger society. Of particular importance in their analysis are the systems of education and industrial relations. Moreover, these interact in coherent, self-reinforcing systems that are rooted in the distinctive histories of specific societies. The recognition of this "societal effect" leads to an important concluding point about comparisons between societies. In their words (Maurice et al., 1984: 263):

Our study of the "historical development of the agents" indicates that analytical categories must themselves be socialized. The French foreman cannot really be compared with the German *Meister*, nor the top-level manager with the *Leitende Angestellte*, because categories of agents were constituted in different social spaces . . . Each institution only makes sense in connection to the nature of social relations on which it is based. In this sense our approach tends to *conceptualize differences rather than similarities* by pointing up the non-comparable quality of the analytical categories.

In the chapters that follow, I try to show the links between the social and political attitudes of French engineers and the unique French institutions that have shaped their market situations. At one level, the book investigates the effects of different labor market positions on the interests and outlooks of individuals. That analysis relies on a basic classification of the interviewees into four career-based categories: *techniciens supérieurs*; *ingénieurs autodidactes*, *ingénieurs diplomés*, and *chefs de service*. These categories are explained in the description of the sample in chapter 2.

At a second level, the book attempts to specify the ways in which the organization of technical careers in France is distinctively French. That analysis makes some use of the sociological literature on engineers in other countries, but only to the extent necessary for identifying largely "noncomparable" French patterns. In the process, it argues for a kind of comparative sociological research that focuses on mesosociological differences among nations rather than such macrosociological differences as national culture. Many studies that emphasize the unique aspects of French society have tended towards one form or another of cultural determinism (Almond and Verba, 1963; Crozier, 1964; Hoffmann et al., 1963; Hofstede, 1980). The French institutions emphasized in this book are not global character-istics of French society any more than they are characteristics of the work done by French engineers. Rather, they express the social definition and organization of technical workers within French society.

Finally, this study attempts to specify the institutions that shape the organization of French technical careers, the processes by which they do so, and to a lesser extent the unique historical conditions that gave rise to them.

A case study of engineers in two firms in one country, this investigation does not pretend to "test" class theories or to "demonstrate" the character and significance of national differences. Nonetheless, by simultaneously examining attitudes, work situations, career structures, and the French institutional context in the light of enduring debates about the new middle class and the logics of industrialism and capitalism, it fills gaps in our knowledge and advances new lines of inquiry.

2

The companies: PAMPCO and TELECO

The theories discussed in chapter 1 – theories of professionalization, proletarianization, and a new working class, are theories of social change. In each case the argument is that science-based industries give rise to important changes in both the organization of technical work and the ideologies of technical workers. Empirical investigation of these claims must therefore examine not only individual attitudes, but also the structure of work and of work organizations. Thus, this study is limited to engineers in two firms, an old metal-working company and an electronic–telecommunications firm. This chapter explains the research strategy, examines the companies and research sites, and describes the samples of engineers interviewed.

The focus on French engineers

Neither the neo-Marxist nor the non-Marxist theorists of post-industrial society dwell on any particular occupation as the leading edge of the predicted changes in work and ideology. Rather, they speak in quite broad terms of professionals, knowledge-workers, experts, and technologists. Moreover, there is far from complete overlap in the sets of occupations mentioned as characteristic of what Bell calls "the professional class" and Mallet "the new working class."[1] Nevertheless, there is a large overlap centering on the "technical employees" in the "science-based industries." Engineers are particularly well suited to an investigation of the post-industrial society theses, for as workers whose primary task is to mediate between science and production, they should be especially sensitive to significant technological developments.

French engineers are of particular interest, both for the reasons mentioned in chapter 1, and because recent studies of American (Zussman, 1985) and British (Whalley, 1986) engineers create opportunities for studies of yet other industrialized societies to identify distinctively national characteristics.

The choice of metal-working and electronics

We have been using the phrases "old industry" and "new" or "advanced industry" as though work organizations varied according to some self-evident and unproblematical scale of modernity. In fact, the concept of technology is fraught with difficulties. First, it is not unidimensional, and different researchers have focused on different dimensions. Some emphasize the nature of the materials being worked on, some the state of the knowledge that underlies the transformation process, and some the characteristics of the operations or techniques used to perform the work.

Second, it is not always evident which if any category of a dimension of technology is most "modern." Continuous-process production may be more technically complex than assembly-line and large-batch "mass-production," and it may also be a more recent arrival on the industrial landscape, but such great growth industries of the post-World War II era as computers and aerospace show little potential for transcending their current forms of traditional unit and small-batch production techniques. As Reynaud (1972: 534) notes, operations technologies must be suited to the characteristics of particular products, and new products are constantly coming on to the market. Many of them are too complex, customized or rapidly evolving to be profitably produced *en masse*, much less by continuous-process techniques. Since the product mix at any one time is a function of widely varying supply and demand factors, there seems no way to predict what sorts of product mixes and associated operations technologies will prevail in the future.

Nevertheless, it is possible to identify certain technological features that are shared by most of the new and rapidly growing industries, that distinguish them from older industries, and that speak to the theoretical issues under examination. What stands out about such post-war growth industries in manufacturing as aerospace, electronics, telecommunications, plastics and synthetic fibers, is the importance of scientific research and technical personnel. Moreover, it is this science-based feature of these industries that is regarded as sociologically significant by the various theorists of post-industrial society. In its most extreme formulations, knowledge is viewed as becoming the product itself and technical personnel are thought to be replacing manual workers.

The science-based industries vary widely in their operations technologies, but they are all high in "knowledge technology," the amount and scientific sophistication of the knowledge needed to design and produce a particular product. Convenient measures of knowledge technology are the proportion of a work force that is technically qualified or the proportion of a product division's operating budget devoted to research and development.

There remains the issue of the environmental aspects of organizations, but it raises few difficulties. Although there is some debate about the evolution of organizational environments, "most observers argue that the environments of organizations are becoming more complex and uncertain over time" (Scott, 1981: 178). Moreover, science-based industries face particularly uncertain and unstable market conditions because of their higher rates of product innovation. Burns and Stalker (1961) offer considerable evidence on this point with regard to electronics firms. The fact that organizational structures usually represent adaptations to environment as well as technology creates no problem, for complexity and uncertainty are also characteristic of advanced technologies (Scott, 1981: 212–222). In short, the environment here being called "modern" reinforces the consequences for organizational structure – and thus work experience – of "modern" technology. Yet, since not all science-based industries face such complex and unstable environments, these characteristics of the environment must be regarded as independent criteria for selection of both an industry and a firm.

The ideal "advanced" industrial setting is in an industry that is new, "science-based," technically uncertain (experiencing rapid product innovation, for example), characterized by "reciprocal interdependence" (Scott, 1981: 211–212) in work-flows, and subject to considerable market instability. The traditional setting is in an industry that is old, non-science-based, characterized by pooled or sequential interdependence in the work-flows, and stable in both technology and environment. Electronics and metalworking fit these criteria well. The metal-working industries range from basic iron and steel processing to the manufacturing of a wide variety of cast, forged or assembled metal products, but they share a metal-working technology developed in the nineteenth century and little changed since then, despite considerable mechanization of the more routine tasks. They employ engineers – indeed, the emergence of the engineering profession is closely associated with the rise of the metal-working industries in the nineteenth century – but engineers account for a small proportion of their work forces. And basic research accounts for a tiny fraction of the work force and relevant budgets.

Electronics also encompasses a wide variety of products, ranging from such mass consumption items as calculators and digital watches to such sophisticated and often customized professional products as computers and advanced telecommunications equipment. What characterizes the industry as a whole is not simply its high position on knowledge technology measures, but its technical uncertainty, reciprocal interdependence, and unstable environment. The technical uncertainty stems from the rapid pace of technological innovation that has seen "generations" of electronic components – tubes, transistors, chips, etc. emerge in rapid succession since World War II and quickly render their predecessors obsolete. The degree of reciprocal interdependence varies somewhat from product to product in electronics, but wherever the product is complex and innovation is on-going and rapid (stereos, computers, electronic telephone exchanges) or the product is highly customized (simulators), the designers and producers of one subunit must be communicating their own requirements to all other affected groups and at the same time taking the changing needs of those groups into account in their own technical decision-making.

However, conditions vary widely within such broadly defined industries. Even in the case of specific products, different firms and different factories within the same firm may be more or less technologically advanced. Thus, it is important to choose appropriate firms and research sites within them.

The traditional firm: PAMPCO

PAMPCO, a pseudonym for the pipe and metal products company, was founded in 1856. In the 1870s PAMPCO began specializing in the production of iron pipes and related foundry pieces (valves, fittings, manhole covers, etc.), the products for which it is still best known. By the turn of the century, PAMPCO had grown into a large vertically integrated mining, manufacturing, transport and pipe-laying enterprise, and had expanded its sales into such foreign markets as Brazil and Japan.

By the 1970s, PAMPCO was producing cement, steel and plastic pipes as well as iron ones. It was also manufacturing other metal products, ranging from cast iron engine blocks to a variety of industrial valves. Although gradually losing ground to these other products, pipes continued to account for roughly 60 percent of sales revenues, with iron pipes still generating two-thirds of the pipe revenues. More than one-half of this iron pipe is produced for export to foreign markets. Except for the Research Center that served the entire company, the research for this study of PAMPCO was confined to the division of PAMPCO that makes iron pipes and related foundry pieces.

The technology for producing these products has evolved slowly during the past century, and hardly changed at all since the early 1950s. The two major breakthroughs, the process for casting pipes in spinning molds and the process for making iron less brittle through the fusing of magnesium with graphite, were invented in 1914 and 1948 respectively. Most of the other improvements in productivity and quality derive from refinements in traditional techniques and from the mechanization of production.

One other measure of PAMPCO's knowledge technology is revealing. *Cadres techniques* (engineers) constitute (in 1977) 3.2 percent of the roughly 12,000 employees, and technicians and draftsmen another 9 percent. The

3.2 percent figure closely resembles the quasi-official figure of 3.3 percent for French forges and foundries in general (UIMM, 1977: 13, 25–26). By contrast, the electronics firm chosen for the research employed 9,600 persons in 1978, of which 1,300 were *cadres* and 2,535 *techniciens*. No figures are available on the exact percentage of the *cadres* that were *cadres techniques*, but a conservative estimate would be 75 percent.[2] The resulting percentages of *cadres techniques* and *techniciens* are 10.2 and 27, three times the corresponding percentages at PAMPCO.

Regarding the environment in which PAMPCO operates, the outstanding features concern the market for pipes and related foundry pieces. First, the market is international in scope; foreign sales account for more than 35 percent of the revenues from cast iron pipes (*Informations Economiques*, 1977). Second, PAMPCO enjoys a near monopoly position on the French market and a strong status in the world market. Nevertheless, foreign producers are becoming increasingly competitive at a time when the French demand for PAMPCO's products has been gradually receding. Thus, after decades of steady growth, the company experienced declining sales in 1976 and 1977, and suffered a net operating loss in 1977.

If PAMPCO is a traditional French metal-products company in its technology and environment, it is even more so in its organizational structure and management style, which are both bureaucratic and paternalistic. This is true despite the fact that since 1970 PAMPCO has been part of a major, Paris-based conglomerate, SAG–PAMP. As one of the two main companies in the larger group, PAMPCO maintains its former identity, its longtime headquarters in the Lorraine, and considerable autonomy. Within the company, however, administration is quite centralized and formalized. Operating responsibilities are spelled out in detail in individual job descriptions; and output and cost information are collected daily, computer-analyzed at higher headquarters, and evaluated by management in terms of pre-designated objectives. Yet the current chief executive "still thinks this is the nineteenth century, runs the company as a personal business rather than as a group project," and insists on having a voice in virtually all substantial decisions, to quote one executive. Similarly, factory directors still engage in a fair amount of personal supervision, intervention and "leadership." The local director of the 1,800 person production site where this research was done made daily inspection tours of the factory, involved himself in every aspect of operations, upbraided engineers he caught not wearing a safety helmet, and monitored the behavior of his *cadres* off the job as well as on it.

There are in effect three separate organizations at this original site of the company. First, there is the factory, the oldest, largest and, according to executives at company headquarters, the most traditional of PAMPCO's plants. It includes a large coke-processing center, blast furnaces, pipe spinning and cooling buildings, a building for producing related foundry pieces, a power station, and various supply and maintenance shops. This 1,800 person factory operates twenty-four hours a day, and the production and maintenance engineers working there complain about the sixty-hour work week and about getting called out from their local, company-owned homes at night and on holidays.

Located near the factory's administrative offices are the Division's Services techniques, a group of about 100 engineers, technicians and draftsmen who are responsible for planning, costing, and designing improvements in work organization and equipment. Since the factory next door is the largest in the Iron Division of the Pipe Branch, the Services techniques' engineers tend to work on projects related to it. Nevertheless, they and their local director report to functional superiors, and complain of the difficult relations with the factory and the cumbersome centralization.

Finally, there is the Centre de Recherches, officially now a separate company within the SAG–PAMP group, but nevertheless responsible to the President of PAMPCO. The Centre de Recherches is housed in a pleasant group of modern buildings a few hundred yards away from the grimy industrial complex of the factory. There, roughly 200 engineers and technicians work at improving the performance capabilities of PAMPCO's materials, products and production processes. The technical employees work in teams that specialize either in specific product lines or specific materials (metallurgy, chemistry, plastics, etc.). The research they do is quite applied, and would be called development in many firms.

In summary, this PAMPCO site well represents traditional French metal-working. The products and the techniques for producing them are old, the ratios of research and technical workers to all personnel are low, the organizational structure of the firm is quite bureaucratic, and yet the style of management is fairly paternalistic.

The science-based firm: TELECO

TELECO is the pseudonym for an electronics and telecommunications company best known for its production and installation of telephone exchanges (automatic switching centers), but that also produces electronic components, professional radar and radio equipment, and electronic simulators (i.e. flight simulation cabins for pilot training).

The technology for producing such equipment is highly advanced. As indicated earlier, the proportion of TELECO employees who are technically trained is much higher than at PAMPCO. It is more difficult to compare expenditures on R&D, for the revealing reason that at TELECO most research is not clearly differentiated from production. To quote a company brochure: "We have kept the expenditure necessary for research and development high. These activities are decentralized, since more than

90 percent of research work takes place in production plants" (TELECO: 1976: 17).

As to technical uncertainty, the telecommunications industry is noteworthy for its pace of product innovation. Since 1962, there have been three major revolutions in the technology of telephone switching: the change from electromagnetic systems to the "crossbar system" in the 1960s, the emergence of computer-controlled electronic exchanges in the early 1970s, and the very recent development of digital time-division switching. As for electronic simulators, each one is a new product, tailored to the airplane or vehicle being simulated, and produced in extremely small batches. This is not a matter of applying well-mastered principles to customized designing, or even simply incorporating the latest advances in electronic components in increasingly sophisticated products. It also involves the ongoing development of new simulation capabilities, such as high definition, closed-circuit visual systems. This is one reason that simulator design and production requires "reciprocal interdependence."

The rapid pace of technological change yields market uncertainty, even for the industry's leaders, for today's loser can quickly become tomorrow's winner on the strength of a product or processing innovation. TELECO recently lost a major telephone exchange contract with the French government, after investing millions in developing a prototype. Rapid product change and product complexity also complicate the procurement of inputs, yielding a more complex environment with regard to suppliers. PAMPCO can rely on a few suppliers of basic ingredients – iron ore, coal, magnesium, etc. The mix of ingredients that TELECO purchases is more diverse and changing.

The marketing of such scientifically sophisticated products also entails complex relations with customers. TELECO not only makes, sells, and installs high technology products, it also services them and trains customers in their use and maintenance. As with PAMPCO, the market is a professional rather than mass-consumer one, and the customer is frequently some local, national or foreign government office. With TELECO, however, the relationship with these customers is complicated by the fact that the government's experts are involved in the entire production cycle, from funding research and determining performance specifications to testing prototypes and designing modifications. Issues of military secrecy and patent rights add to the complexity and uncertainty. Finally, in the case of simulators, the market itself is ill-defined and changing, as new uses for simulation are discovered, costs reduced, and new models developed.

Nevertheless, TELECO has enjoyed robust growth: sales increased by almost twofold between 1972 and 1977, net earnings doubled. Although less impressive after discounting for inflation, TELECO's real growth stands in marked contrast to PAMPCO's stagnation in the mid-1970s. One beneficial consequence for employees is expansion in company employment, and hence promotion opportunities. In the five years from 1973 to 1977, TELECO increased its total work force from 9,385 to 9,600, despite selling one small division. During the same period, PAMPCO's total work force declined from 12,815 to 12,130.

With regard to organizational structure, TELECO is less centralized and formally organized than PAMPCO, and is characterized by a mutual adjustment mode of coordination. Indicative of the relative decentralization is the autonomy enjoyed by the product divisions. Although composed of only 500 employees, the Simulator Division does its own hiring, including that of *cadres* (unlike PAMPCO's pipe division), manages its own sales as well as development and production, and makes its own product decisions. The lower degree of formalization is evident from the lack of job descriptions at TELECO. Coordination by mutual adjustment is assured by the powerful role of project coordinators and the frequent use of task forces. Within divisions, lateral contacts are facilitated by TELECO's "matrix organization," a structure that permits "the simultaneous operation of vertical and lateral channels of information and of authority," with functional departments serving as "home bases" and project groups coordinating the work of selected functional specialists (Scott, 1981: 220).

Nevertheless, TELECO remains bureaucratic in many respects. As Zysman (1977: 134) notes, TELECO's parent company is a traditional French firm characterized by "centralized authority . . . sharp divisions between strata and among operating units, conservative and cautious management at the lower levels . . . and formal procedures that are often blamed for its sluggish performance." One sales engineer who had previously worked for IBM complained bitterly about how much more bureaucratic TELECO was. A personnel officer agreed, and suggested that one reason was that TELECO sold large, expensive products to big, centralized organizations, such as French ministries, and that buyers usually prefer to deal with representatives of at least equal status.

Scholars such as Michel Crozier (1964) and David Granick (1972: 46) offer a more comprehensive explanation for why, in Granick's words, "all French company planning and decision-making in general is highly centralized," and subsequent chapters will address this issue in more detail. The point here is simply that while TELECO's organizational structure differs in significant ways from that of PAMPCO's, it is still far more bureaucratic than the "organic" model that Burns and Stalker (1961) view as natural to electronics firms.

The research and interviews at TELECO were carried out at three separate sites. The first is the Simulator Division plant, located in the recently built *zone industrielle* of a small suburban town west of Paris. Of the more than 500 persons working there at the time, about 200 were engineers (including 18 in sales), 200 technicians, 70 workers and foremen, and 55 clerical workers. These numbers were growing steadily. It was difficult to make an appointment with the local personnel director because he was so busy interviewing candidates for new positions. In many ways the Simulator Division resembles a small engineering firm as much as the electronics products division of a manufacturing company. In particular, it has few production personnel and an exceptionally high proportion of engineers to workers. Consequently, research for this study was also carried out at two sites within TELECO's Telephone Division: a provincial factory for the production of electronic telephone exchanges, and certain Telephone Division R&D departments at company headquarters in Paris.

The provincial factory, located on the edge of a regional city in the west, employed about 840 persons, of which roughly 100 were engineers and 500 workers (mostly women) and foremen. Most of the engineers worked at desks in large multi-engineer offices, and spent the bulk of their time calculating costs or ironing out small problems in circuitry designs. At the company headquarters in Paris were over 2,600 employees, 750 of whom worked in the Telephone Division. Of these, roughly 300 were engineers and another 350 were technicians.

Taken together these three TELECO sites represent the advanced industrial setting. The complex, highly engineered products represent recent and rapidly evolving developments in science and technology, the ratio of technical workers to all personnel is extremely high, and production involves a good deal of reciprocal interdependence. The environment is complex and unstable, although perhaps more stable than in the case of consumer electronics. Finally, the matrix organizational structure is less centralized and formalized than at PAMPCO, but still fairly bureaucratic and hierarchical in important respects.

The samples, data collection and data analysis

Many corporate executives are engineers by virtue of their educational credentials, but no longer do any technical work. This study excludes them, focusing on those persons who do engineering for a living. Engineering means the application of scientific knowledge to production. For the purposes of this research then, an engineering diploma is neither a necessary nor sufficient criterion for inclusion in the sampling population. On the other hand, scientific knowledge can be applied only if it is first possessed, and given the similarity in the work of many engineers and technicians at early stages of an engineer's career, educational credentials are useful for distinguishing engineers from technicians.

The situation is complicated in France by the sharp distinction between *cadres* and non-*cadres* in industry. It is important to bear in mind that almost all graduates of recognized engineering schools in France – *ingénieurs diplomés* – begin their careers as first level *cadres*. If they take jobs that make use of their technical skills, they are *cadres techniques* – technical officers – as opposed to *cadres administratifs*. This *cadre* status is an official one recognized by French law and entitling its bearers to certain pension rights and other privileges.

Some *cadres*, however, are former workers and technicians who have risen through the ranks, being promoted to *cadre* status on the strength of demonstrated ability and loyalty. To distinguish them from *cadres diplomés*, they are often called *cadres autodidactes* (self-taught), "a category which to all intents and purposes does not exist in Germany" (Maurice et al., 1984: 253). In the case of *cadres techniques*, the distinction is often between *ingénieur diplomé* and *ingénieur autodidacte* (or *ingénieur maison*). The presumption is that all *cadres techniques*, whether *diplomés* or not, possess a high level of technical knowledge through on-the-job experience if not formal schooling, but for inclusion in the sample population, we have required evidence of some formal technical training. In fact, many *ingénieurs autodidactes* have at least a *Bac Technique* or the equivalent in France's credentials system. The *Bac* is France's highest secondary school diploma.

Included in the sample population are also those *techniciens supérieurs* who have at least a *Bac Technique*, for their technical responsibilities resemble those of an engineer by American standards. Being neither *cadres* nor *diplomés*, they are in a far different career position than young *ingénieurs diplomés*. Many eventually become *cadres* but have virtually no chance of rising to positions of higher management, while the young *ingénieurs diplomés* have a good chance of doing so. Nevertheless, *techniciens supérieurs* do high level technical work and often command subordinates. For example, the large team of technicians that tested assembled telephone exchanges at TELECO's regional factory was commanded by a *technicien supérieur* who had two years of post-*Bac* technical training and who reported directly to the *chef de service* or department head. Moreover, in view of the research interest in developments in the division of labor and authority within the technical stratum, it is desirable to have the sample cover the full hierarchical range of what are, functionally speaking, practicing engineers.

Differences in the situations and attitudes of different levels of engineers – *techniciens supérieurs*, *cadres autodidactes*, and *ingénieurs diplomés* – should throw light on the meaning of these situations and attitudes. Finally, to contribute to comparative perspectives, it is desirable to include *cadres* and non-*cadres*, for by doing so we can shed more light on the significance of the *cadre* status, a status particular to and very significant in the French division of labor and authority. Therefore, we have reached down into the category the French call *techniciens*. At the same time, we have included only the highest level of *techniciens*, thus excluding those regarded as technicians in the United States.

A similar problem of definition arises at the upper end of the engineering hierarchy. At what level in the hierarchy does an engineer stop being a practicing engineer and become a manager? Moreover, because many engineers do rise to higher management positions in the course of their careers, it is desirable to include the first level of technical management (in the American sense of the word) in the sample of engineers. The sample thus distinguishes a fourth and highest level of engineer, that of *chef de service*, or department head. These *chefs* are all *ingénieurs diplomés* who have risen to the first level of management in the sense of officially commanding the basic unit or organization rather than simply individual assistants or an informal team. In the nationally formalized French system of industrial rankings, they are usually *cadres IIIA* or *IIIB* rather than *cadres I* or *II*. Thus, the sample population is defined to include all *techniciens supérieurs*, *ingénieurs autodidactes*, *ingénieurs diplomés*, and *chefs de service* whose primary work involves the application of scientific knowledge to industrial production.

So far we have distinguished engineers with regard to their position in the vertical, but not the horizontal division of labor. Essentially the issue here is one of technical versus non-technical work rather than of formal or informal categories in the technical division of labor. Thus, we include sales engineers at TELECO and not at PAMPCO because only at PAMPCO is a career in sales an alternative to one in engineering, while at TELECO a position in sales represents success in an engineering career. More generally speaking, we have defined the range of technical functions to cover all technical work from conception to production, from the most theoretical – R&D – to the most applied – maintenance. In accordance with the practice of French industrial associations, we have distinguished five major categories along this continuum: (1) R&D, (2) Design, (3) Methods, (4) Production and Maintenance, and (5) Sales. The very few engineers doing technical work that did not fall into one of these five categories – technical authors, for example, have been excluded from the sample population.

The sample also includes fifteen *cadres administratifs*, all of them at TELECO. These are personnel managers, financial administrators, and members of the legal staff. They serve as a comparison group for the purpose of assessing the relative significance of occupation and stratum. They are particularly relevant to the argument about a "service class."

It is important to have a substantial number of *ingénieurs diplomés* in the sample, for this is the central category of engineers in France. The *chefs* represent *ingénieurs diplomés* at a later stage in their careers, but the *techniciens supérieurs* and *ingénieurs autodidactes* are included as much for perspective on the *ingénieurs diplomés* as for their own sake. Thus, the final sample design includes almost twice as many *ingénieurs diplomés* as it does members of any other level. Moreover, these are somewhat concentrated in certain categories (company and function) in order to permit some controlled explorations of differences among them. The sample obtained closely resembles the sample finally sought, and is presented in Table 2.1.

In order to focus on the significance of old–new firm differences in a sample that is too small to permit many statistical controls, the samples in each company were designed to be as closely matched as possible along other dimensions that seemed likely to affect an engineer's ideology. Thus, the distributions of levels and functions are fairly similar in the two companies. For levels, the sample obtained emerges with exactly the same proportions at each company: 23 percent *techniciens supérieurs*, 19 percent *ingénieurs autodidactes*, 39 percent *ingénieurs diplomés*, and 19 percent *chefs de service*. For technical functions, the breakdowns are not exactly parallel, but after appropriate reclassification the ratios of conceptual to applied positions at the two companies are very similar, 54:46 at the old versus 57:43 at the new. For one other variable, age, an effort was made to match the company averages and distributions. In fact, the new firm sample turns out to be slightly younger than the old firm sample, 34 on average as opposed to 36. With all these variables, a major principle was homogeneity within the basic types as well as matched and balanced distributions among them. That is, limited by the small numbers, the research focuses on homogeneous clusters of typical types of engineers.[3] It is noteworthy that the sample is limited to white men of French nationality, but that is typical of French engineers and *cadres*, especially in private industry (Boltanski, 1987: 190–196).

The interviews themselves were conducted at the research sites in offices provided for that purpose. They ranged from two to five hours in length, but averaged about three and a half hours each.[4] When the appointments for interviews were made, the interviewees were given a short written questionnaire and asked to fill it out by the time of the interview. This enabled us to obtain some of the simplest objective data on such things as age, educational background, job history, current work, and family situation, without wasting interview time.

The interviews, done in 1977–78, consisted of a semi-structured series of pre-tested, open-ended questions on the following topics: (1) social origins; (2) educational and job history, including the reasons for choosing engineering as an occupation, changing jobs if ever, etc.; (3) career evaluations, hopes, and worries; (4) perceptions and evaluations of the existing divisions of labor and authority at work; (5) family and home situations, including leisure activities and aspirations for children; (6) community involvement and patterns of sociability; (7) voting history and political participation; and (8) images of class structure.[5]

Table 2.1. *Sample obtained*

PAMPCO ("old industry")

	Function				
Level	R&D	Design	Methods	Production and maintenance	Total
Chefs de service	2	3	1	4	10
Ingénieurs diplomés	6	6	3	5	20
Ingénieurs autodidactes	0	3	3	4	10
Techniciens supérieurs	4	4	0	4	12
Totals	12	16	7	17	52

TELECO ("new industry")

	Function						
Level	Etudes	Methods	Production and maintenance	Sub total	Sales	Administration	Total
Chefs de service	5	3	3	11	2		13
Ingénieurs diplomés	14	4	4	22	6	12[a]	40
Ingénieurs autodidactes	7	2	2	11	0	3[a]	14
Techniciens supérieurs	6	3	4	13	0		13
Totals	32	12	13	57	8	15	80

[a] The *cadres* in this column are not *ingénieurs* – i.e. *cadres techniques*, but rather *cadres administratifs*.

It is important to note that the formal interviewing was accompanied by a considerable amount of observation and other field research at both the individual and organizational levels. At the individual level I spent entire mornings or afternoons accompanying engineers on their rounds, attending staff meetings, or simply observing a room full of design engineers and technicians. I also ate a great many lunches in plant cafeterias and a good many dinners with interviewees and their families in their homes. At the organizational level, the research involved learning a good deal about the following:

the history of the firm and plant; the products currently produced and the markets for them; the research and production processes; formal and informal organizational structures; the recruitment, classification, remuneration, evaluation, and promotion of technical personnel; the decision-making and supervisory processes; and the labor relations situation. Obtaining such information was possible only because the company personnel officers and their staffs, especially at PAMPCO, were extremely cooperative.

3

Pierre in his own words

Most of this book's argument revolves around the variable analysis of aggregates of engineers. To make such analysis more meaningful, this chapter provides a glimpse of the actual people behind the percentages and quotes to come. It does this by presenting most of one engineer's own account of his work and career experiences. A few minor biographical details have been altered to protect the identify of the interviewee.

At the time of the interview, "Pierre" was a 32-year-old development engineer at TELECO. According to a questionnaire filled out before the interview, he comes from a family of farmers on his mother's side, skilled and lower white-collar workers on his father's side, and grew up in a small village (250 inhabitants) in southeastern France. Although his father, a stonemason, had not been educated beyond primary school, Pierre completed his *Bac* (France's highest secondary school diploma), and went on to obtain an engineering diploma from a modest engineering school. After two years of military service and two short-lived jobs at other companies, he came to TELECO as an *Ingénieur* I.

After six years in the same department, he is an *Ingénieur* IIIA in charge of a team of ten engineers and technicians. He estimates that he works about forty-two hours per week, 95 percent of it in his office, the other five percent in the laboratory, and says that he rarely takes work home with him. Home is a house that he and his wife own in a small town in the Paris suburbs. His wife has a master's degree, and works full time as a technician. Married seven years, they have one small child.

What follows is a slightly edited and abridged version of the *verbatim* transcription and translation of the interview with Pierre.[1] The interviewer's questions are included only where needed to make the respondent's replies intelligible.

[Interviewer: To begin with, could you briefly describe for me your current job?]

"For the past year and a half, I have been responsible for a group involved in hardware conceptualization on the XTE project. It consists in defining and realizing, up to the point of workable diagrams, one hardware section of the exchange. There are three such teams, each one quite specialized in its own field. When I talk about realizing such a project, I mean writing out all the specifications and producing the diagrams according to these specifications. Our own responsibilities end at this point; other teams are responsible for the actual production of the printed circuits, etc. Thus, our team is only involved in the conceptual aspects of designing hardware.

I have a team of five engineers and five technicians. My role is to divide the work among them, to participate with each engineer in the definition of the part he has to realize, and then to supervise the work. My supervision is done at the level of general conceptualization to check if the work is in accordance with the specifications that we hope to achieve.

Yes, most of the team members use the familiar form of address with each other, even between the technicians and engineers. There may be a few exceptions, but they are rare. As a matter of fact, this goes for me too, in my relations with them. The team involves a nucleus of members who have worked together for the past five or six years. New members, including two engineers, have joined in the past year and a half, and they have integrated quite well. We had to hire more people when the project became definite and we got a firm order. As a result, the team we had for the development of the models was expanded; we had to hire more staff. When I say 'hire,' I mean either from within or outside the company.

When I speak of supervising, it often means participating in the design work. When there is a problem, one has to participate in the solution of that problem. Once we agree on the path to take to resolve the problem, the engineer follows through by himself, and I move on to another.

Yes, we have frequent contacts with the other teams working on the same project, but not so much with other departments. However, we do rely on other departments for specific purposes. For example, there exists in our company a department which is in charge of all problems concerning components. We therefore call on them when we have such a problem, but our contacts with them are brief, an hour or so. Sometimes they call on us when they have a problem. Since they are responsible for buying and checking out the circuits which we use in our project, obviously there are exchanges between the project's technical teams and the technical components department.

Such division of the project into separate teams means that at my level, one is very specialized. One is less so at a higher level.

Yes, I would be able to head up another team, at least one of them. However, one of the other teams is even more specialized than we are. Therefore, I don't think that I would have the necessary technical knowledge to head up such a team. As for the other one, I am confident that I would not have any problems after a minimum period of adaptation. As a matter of fact, I am familiar with their problems. Each team is specialized in its own field, there is no question about that, but all the documents that each team produces are distributed among the other teams. As a result, one can keep in touch with what the others are doing. Unfortunately, one does not always have the time to do so. But, besides, we use the same products to do different things. So all in all it's not as different as it may sound. All in all, specialization is not such a disadvantage or hindrance: people have a good knowledge of their subject and are therefore able to react and move on quickly when faced with a problem. The one disadvantage which specialization presents is at the level of conceptualization. After a certain time, one may not be able to come up with new ideas. One has such a deep knowledge of the matter at hand that it's hard to find a new angle and take a step forward . . . Indeed, if people were switched from one team to another, there would very likely be more creativity, more new ideas. But the practical aspect of running a business imposes imperatives, and that's why there is not more rotation. I must admit that at certain times I do feel a bit restricted. I'd like more variety; working on the same subject, even if a vast one, can become tiresome.

The knowledge I use? I use very little of what I learned in school, as far as the specifics of my studies are concerned. However, as far as the general technical education I received, I do make use of it. It's useful for me to know pure math and physics, for example, but I can't say that I am really using that knowledge. All I use from my courses is based on the binary code, and one certainly does not need to study until the age of twenty-three for that. Still, school has given me a solid core of knowledge and a frame of reference which help me to resolve problems when they arise. And my situation is somewhat specific to the telephone industry. If I were working for a company involved in semiconductors, for example, I would be using my knowledge of math, physics, and even chemistry. At the conceptual level in the telephone industry, one does not need to know for example how a transistor works, due to the type of integrated-logic circuits used. I am given a specification; my circuit is a black box, and I am told what to put in and what is expected to come out of it. We have reached the point where we set up series of boxes together in order to realize a global function, but we are seldom obliged to formulate any thought at the inception level, and we are not concerned about the physics phenomenon itself. My job involves electronics, but at the level of general conception, and not at the level of a primary product which requires knowledge of the laws of physics. Nevertheless, I do feel it was necessary for all of us here to pursue the type of studies we did, because it allowed us to orient ourselves in the different domains of technical work.

I am aware that I have learned a lot through work experience. The technology has evolved so rapidly in the past ten years that I have had to learn things which did not even exist at the time I was in school. I have learned through in-house training and the literature of our suppliers. I have had the satisfaction of having grown professionally, to have acquired new knowledge. On the other hand, I do not feel that I am using all that new knowledge. As I mentioned before, every job entails certain imperatives that are restrictive. However, I am sure that I have used that knowledge with the imperatives of the projects in which I have been involved.

When a technical problem for which I lack sufficient knowledge arises, I first go to my colleagues, because it is easier and faster . . . If I feel this problem is not an entirely new one, I ask around among the people who have been using the same product. Even if they are not familiar with that specific problem, they can contribute their own experience and point of view. That way you gain insight and a better understanding of the problem. As far as the literature is concerned, you can never find the piece of information that will give the exact answer. Besides, if it did exist, our company does not have a library service with an indexed catalogue of all aspects of our technology. In fact, our only resources are the documents which circulate, and we therefore probe each others' memories. Any new product is introduced through literature from the manufacturer. After analyzing the documentation, if we still have questions or problems, we go to the manufacturer, who has teams of experts equipped to answer our inquiries . . . Well, these experts are more business than technically oriented, but within the manufacturing company we usually end up finding someone who has the technical know-how. If not, we may experiment to see where we stand . . . No, we don't need authorization from a superior to contact outside experts; it is automatic.

[At this point, the interviewer shifted from questions about current work to questions about career, beginning with one about the initial decision to go into engineering.]

No, it wasn't in the family. It was more a matter of circumstances than vocational calling . . . Since I had no idea of what I wanted to do schoolwise, I went to the village school up to the end of the required program [rather than to a school in the city 7 miles away]. Since I did rather well in that school and since my parents were in favor of my pursuing my studies further, I went to the city junior high school in eighth grade and enrolled in a vocational program. This was due to the fact that at my previous school I had not followed the typical track that usually leads to the *Baccalauréat*, and had not taken any foreign languages. After one year there, my grades were good enough for me to switch into the regular program within a technical section . . . During the year of my *Bac*, a decision had to be taken. At that point I still didn't have any particular goal and didn't know what I wanted to do, except that I knew that I was interested in the technical field. The engineering schools were out of the question as far as I was concerned, because they entailed two years of preparatory schooling, where competition is extremely stiff and the final outcome of the entrance exams for engineering school far from certain. At one point I heard about an alternative program, which immediately attracted me because admission was gained through the *Bac* and one's high school record. Had that option not been available, I would have gone to university to study for a Bachelor of Science degree, as opposed to an engineering course.

[Interviewer: Do you ever wish you had chosen another profession?]

No, I don't think so. I have had the satisfaction of achieving a goal, even if that goal was not chosen because of a real vocation, a calling.

When I finished engineering school, I had my own idea of what an engineer's job was. Once I actually started working, I found out that the reality was quite different. But the misconception was not entirely my fault; in school they told us that an engineer was supposed to have a number of people working under him, was to be in charge of a project, was over-worked, could never quite keep up with new information and knowledge, etc. In reality, a starting engineer is given a desk somewhere with a pile of documents to peruse and absorb. This goes on for a while . . . then eventually he is entrusted with a project. At the time, the projects were based on circuits which were quite different from the ones we now have: relays, transistors, and such. Therefore, I soon figured out that I was not going to learn much in the technical end of the job. During the year and a half that I worked there [his first company], I had the feeling I was wasting my time, that I was not putting to use what I had learned in school, and that on top of everything I was not learning anything new. Thus, I decided to go see if the grass was greener someplace else.

[Interviewer: Is it different here?]

Yes, because I was lucky enough to join a company that had just started work on a potential project. I was joining a team that was being formed for a future project. From a strictly technical point of view, I was getting involved with new products within a team. It was a conceptual job which entailed working closely with other engineers. As a result of all that, I certainly do not have any regrets about changing jobs. Since then, and with hindsight, I believe that had I been more patient, I would have had the same opportunities with my previous employer. That company has also evolved, but I must admit that I was too impatient. I should have waited, and maybe asked to be moved to another department, but within a company it is always difficult to switch from one department to another.

[Interviewer: How did you get the job at TELECO, through someone you knew?]

No, I sent résumés to a number of companies. My selection of those companies was based mostly on their names, on their image. But I have forgotten to mention one thing. In between those two jobs, I worked for three months at the subsidiary of a big firm where I knew someone . . . I only stayed for three months because they had financial problems and therefore had to reduce their payroll. My departure from them was easy and did not create any problem because there is no binding contract for the first three months, so there was no breach of contract involved in my case. I wasn't particularly pleased, but . . .

[Interviewer: Did you find another job before you were terminated?]

Yes, but I had no choice; I took what was offered to me by the first company that answered my job inquiry. I didn't wait for other offers because then I would have been on unemployment. Rightly or wrongly, I felt that I would be less attractive to any company as an unemployed engineer rather than as an engineer who was looking for a change and new opportunities . . . Had I been unemployed at the time the second company hired me, I would not feel I wasted my time there. But I was gainfully employed in a good company. They approached me and offered me a position; and yet, one and a half months later they told me that I was of no use to them. It is a sign of disorganization, of a lack of sound planning . . . Anyway, I did not worry about it because I was an engineer, I was specialized in electronics, and I knew there was work for me someplace. But of course, I could have faced unemployment for three months maybe . . . No, the company did not help me in my job search. At first I had the feeling that they wanted to calm me down by offering me all kinds of interviews with their subsidiaries, but none of the positions fitted my specialty; it was done just to prove they were trying to help . . . I got a list of replies to my job inquiries, but my present employer was the first to offer me something interesting. So all in all, I did not have too many problems finding a new job.

[Interviewer: Have you changed position or function since coming to TELECO?]

No; the project has evolved, but I have stayed with it all along. My responsibilities have increased, that's all.

[Interviewer: Are there any things that you did in your previous jobs that you now miss?]

Yes, one. In my first job, I had the possibility of doing some field work, of getting out of the office and seeing other people, and of traveling a bit. It's just a personal satisfaction I wish I still had, for we really do not get out enough here. It's strictly on a personal level, not professional. As I said earlier, we do see the end products of our work here, but the atmosphere is different in the field. It's more exciting, it recharges your batteries. It's interesting and helps broaden your mind. You return to the office with new ideas; it's a breath of fresh air. Field work exists here, but other departments are involved in it, those in charge of installation. In the six or more years I've worked here, I haven't gone on site once. What really gives me the most satisfaction is at the level of conceptualization. To figure out how to realize a specific function is quite challenging, because there are so many paths that could be taken. Much of this work involves interaction with others. When you work alone, you quickly reach a point where you rehash the same ideas over and over. It's hard work to be creative in a sort of vacuum. Interaction with others is so much more motivating, ideas bounce back and forth, and you're able to detect very quickly the faults in your own ideas and in those of your collaborators. There is a lot of conceptual work at the beginning of projects, but that does not mean that this activity is not present all along. For instance, even when we reach the stage of realization, we have to review everything in detail and probe more deeply, and that involves conceptualization too.

Most unpleasant? Maybe it's having to exercise control over the team. Controlling does affect your relationships with people, even if you're careful not to criticize and impose your own point of view, but just make suggestions and guide people towards a mutually acceptable way of approaching the problem. Controlling is tedious work, and time-consuming. It has to be done, but it doesn't enrich your experience, and it certainly is not the most interesting part of my job . . . It's just a safety procedure: a careful review step by step of what has been done so as to ensure that nothing has been overlooked or mistakes made.

[Interviewer: How easy would it be to find a new job if you were looking for one today?]

I think it is going to get more and more difficult for several reasons. First, there are fewer job offers at the level I'm at or want to be at, compared to those for entry level engineers. Second, I myself will be more demanding, or have greater requirements than I have in the past. Yet, such jobs do exist; I would just have to bring myself to start looking.

At TELECO, even if the change in ownership brings a reorganization, my position as an electronics engineer in an R&D department seems fairly secure for the next ten years. If any problems should arise, they would affect the personnel in the manufacturing end of the company first. Beyond ten years though, no one can tell, even for those of us in R&D. TELECO at this point in time seems to be well established as a leader in its field, but nothing is certain forever, and it may not last. It is also obvious that the market share which TELECO has had for the past ten years will evolve; it won't be the same. Investments by the P.T.T. [the French postal, telephone, and telegraph system] will most likely decrease and thus directly affect TELECO. Whether the company will then be able to adapt and negotiate within the new market environment, I don't know. As a result, no one is really safe in the long run. The future of the company depends not only on its ability to adapt technologically, but also at the management and organizational levels. Right now, TELECO has a state agency as a principal customer, and is performing quite adequately. When the company has to enter into a more competitive, international market, involving smaller orders and more exports, no one can tell how it will react.

[Interviewer: When you consider the course of your career up to this point, what are your major satisfactions and/or disappointments?]

I feel that the success of a career depends not only on the individual, but also on the circumstances. The latter have been rather favorable to me, and I don't regret any of my moves. As far as my present position goes, I am pleased.

I can't tell you how this evolution compares to my original expectations, for I have forgotten those. I must admit that the scenario that was given to us in school as to the role of an engineer within an organization was not realistic at all. I have had experiences in three companies, and when I compare notes with my classmates, their experiences do not seem to differ from mine. I therefore feel that the expectations generated and that I had before working, were not realistic. What I had not expected at all were the limitations on the role of the engineer. When one thinks about it, those limitations are inherent in the complexity of the organization in which he is working. I was made to believe, however, that an engineer was the head of the company! Perhaps this may have been true twenty-five years ago, but it is not so now. Especially in an entity such as TELECO, where engineers abound. In the technical department here, there must be over a hundred of us, and we're seen as blue-collar workers, not executives. Let's say we are specialized, highly skilled workers, and the technicians are the workers.

Being a *cadre* doesn't mean much; not to me, because I am one. It may still mean something to the technician or the office worker who perhaps view the *cadre* as the person above them who tells them what to do and has certain responsibilities. But as far as I am concerned, I don't view my role as such, and my relationship with them is not based on that. TELECO has undergone some democratization in recent years; yet there still exist remnants of the old hierarchy, such as a special cafeteria for the *cadres*. In fact, democracy is still far away. The proof is that no one here was consulted, and this even includes the general director, when the buy-out took place.

[Interviewer: Regarding the future of your career, what level and kind of position would you like to have attained five or ten years from now?]

My career is not the central factor in my life. I find fulfillment in many other aspects of my life, and therefore do not especially wish to become the head of a department with one hundred people under me. On the other hand, just from the point of view of pride and self-worth, I do not want to stay in my present position forever. I am almost thirty-five years old. There is no question that the next five years will be decisive as far as my career is concerned. The position I'm in when I'm forty or forty-two years old will be the one I'll stay in until the end of my career. Or to put it another way, I do believe that when I have reached that age, it will be extremely difficult for me to market myself at a higher grade outside the company that employs me at that time. My present alternatives then are as follows: either I make a move within five years to pursue a career at a higher grade outside of TELECO, or, if within five years I am still an *Ingénieur* III here, I will look for a similar position in another company out in the provinces, since my family and I would benefit from the quality of life we would find there, as compared to the one we now have in the Paris area. This latter alternative is not my objective; I would rather grow professionally and reach a higher level within my present field. But if it is not to be, then I would want to trade this goal for a life-style out in the country that better answers my needs and those of my family. I think that in any career, the critical point is between age thirty and forty. After that, I do not say one cannot grow or progress, but this progression is not a result of strictly technical abilities. It is then mostly based on the knowledge one has of company politics and internal structure. As far as this aspect is concerned, I really haven't had any experience and do not know how effective I could be.

To be successful as an engineer, it is preferable to be polyvalent [i.e. versatile], at least in the formative years. In that way one doesn't limit oneself, and has a greater scope of choices. But specialization is a necessary evil; you have to specialize in order to be recognized and make a name for yourself. But then again, overspecialization has its own shortcomings, it stunts your career. It is therefore vital to broaden your scope, either by moving within the company or to another company. The latter is even more beneficial, career-wise, for it proves you can adjust not only to the specificity of a new job, but also to the idiosyncrasies of a new organization.

[Interviewer: Has the fact that you graduated from INSA rather than from one of the more renowned engineering schools had a bearing on the advancement of your career?]

It definitely plays a part at the beginning of a career. And now I realize that the entry level job one gets makes all the difference later on in the course of one's career. The gap grows bigger, not smaller. Prejudice against certain schools does exist. Some may be able to overcome it, but rarely. School identification plays an important role. One must admit, however, that those who come out of the Polytechnique or Telecom have already proven themselves. I do not say that they are better than the others, but they have an added asset. Let's take a concrete example with which I am familiar: at TELECO all the engineers who come from Telecom are offered opportunities and advantages that others do not have. When a choice arises, they are undoubtedly the beneficiaries. This is one of the criteria on which a career is based. It is not right, but at the same time, the others have to prove their worth, they have to overcome this barrier through their own performance and job competency.

[Interviewer: I'd like to return to the subject of specialization. Do you think there are any engineering jobs here that are to specialized as to make the engineer's work boring?]

No, there's always something interesting. The problem doesn't lie in how interesting a job is, but how long it will stay interesting. When there's no intellectual challenge, you lose the ingredient that is our trademark, that is, our imagination and creativity. Everything becomes repetitive and routine, and you get in a rut.

Yes, I think you can get out of it. As far as I am concerned, I do believe I can change because I have kept up with new technology and the evolution of the industry. Problems only occur when you're not up to date with technical developments, then it's almost impossible to get out of your particular field.

[Interviewer: Do you think there is a trend towards greater specialization of engineering work in departments such as yours, greater than twenty years ago, for example?]

Yes, but it's inherent in the present technology. Each new product requires such in-depth knowledge and technical know-how that specialization is inevitable. Twenty years ago the engineer was the head of a department. Now, the technical sophistication of the equipment and products requires a greater number of engineers who have become, in effect, technicians. None of their other assets and qualifications, such as management and personnel supervision, are called upon as they were years ago. At least that's been my experience at TELECO, but it may be different at other companies, although I doubt it. In any company that employs a great number of engineers, the latter end up becoming mere specialized technicians in their own fields.

[Interviewer: Some say that this specialization presents advantages, in that it allows one to obtain an in-depth knowledge. Others, however, maintain that routine rapidly sets in and the job soon loses its appeal.]

I rather opt for the second view, although a general overview of many problems does not lead to a knowledge of any one of them. Once an engineer has proven that he can understand a product in depth, that he can resolve problems in a specific field, there is no reason why he should not be able to transfer that knowledge and competence to other fields. The ideal situation therefore would be to utilize the engineer for different tasks in order to stimulate his creativity and sustain his interest. This may be the ideal situation for the engineer, but it may not be for the company which employs him. In a company such as TELECO, people know that an individual is an expert or a specialist in a certain problem, and efforts are made to keep him in that position because it's rather handy when a difficulty arises.

[Interviewer: What about your own job? Has it become routine, or is it sufficiently diversified?]

At the moment I find it diversified enough. I may be concerned with the same type of problems, but I have the opportunity to view them from different angles . . . I do believe that if I objected to working on a particular project, I might be offered an alternative, but we are not presented with different options. When a new project is getting under way, you are asked whether you want to take part in it. That's all.

[Interviewer: Has it ever happened that you could not do a job as well as you would have liked for reasons or time or business considerations?]

Yes, and it leaves you with a feeling of frustration and disappointment at first. But experience makes you wiser. You get satisfaction from seeing a project through to its completion, even if it isn't perfect. It is better to end up with a finished operative product that to try to reach perfection and never achieve it. There always comes a time when you have to stop during the definition of a product and say: this is the way we are going to carry out the project, even if all problems have not been uncovered. Once the realization of a project is under way, you have to face up to reality, be satisfied with the quality level. You have to know when to stop searching and accept the imperfections and restrictions of time and money. The company's objective is to deliver a product. Therefore, there comes a time when conceptualization on paper must be stopped in order to produce a finished product.

[Interviewer: Would it be desirable to expand your decision-making scope; would expanding your range of activities and responsibilities improve the quality of your work?]

Yes, I think so. The limitation on my initiative is due to the hierarchical path I have to follow when a decision has to be made. It's regrettable, in as much as it's time-consuming, restricting, and somewhat reduces my commitment to the product. I'm not saying that the product would be better if I had more freedom of action, but I would be more enthusiastic about my work, and feel more concerned. I believe the quality of my work would improve. I'm not questioning the fact that my superiors need to oversee all decisions at the level of conceptualization and definition of the product, for they have a greater knowledge of the whole project of which my work is only a part. But as soon as we all agree on the definition of what we intend to do, I think I should be given more freedom as far as the actual realization and subsequent management of the project.

[Interviewer: Are you sometimes obliged to work on tasks you don't agree with?]

No, not intrinsically. I believe the rapport with our superiors is such that they do take the time to persuade us that what they are asking for is the best solution. Sometimes we do think that what is being done is not the best solution, or is not what we would have done ourselves under the circumstances, but fundamentally we agree. We are never forced to do something with which we disagree, just because it is an order from a superior . . . It would be impossible to work under those conditions, and our superiors know it as well as we do. Therefore, they take the time to persuade us and to listen to what we have to say. So the final decision represents a consensus. It is not a one-way street, and compromises are often made. Also, it depends on the importance of the matter at hand. Sometimes it entails such a minor detail that it's not worth an argument.

The decision-making process could be different, however. As far as my own work is concerned, there is a need for a person above me with a wider understanding and knowledge of the project. It could be someone who is professionally at the same level as I am; I wouldn't mind. Unquestionably, though, there has to be someone who makes the final decision, just so that it is binding on all and can't be constantly rehashed. As I've said before, we could argue forever in order to try to come closer to perfection, but it is not feasible, or else nothing would ever get done. So at one point a decision has to be made. I do not see the necessity for this decision-maker to be hierarchically superior to me, but that's the way things are set up here.

The control over my work performance varies. In principle, it is rather strict, and variations in degree come from the workload of people in charge of that control or from the importance of the task to be controlled . . . By strict I mean meticulous, too concerned with tiny details. We have to demonstrate, prove again what we've already proved.

[Interviewer: Within the company, do you think there are decisions about which you and other *cadres* should be better informed or consulted on?]

I think *cadres* should be informed on company matters. As far as being consulted and giving our opinions on these matters, I'm not so sure about that. The directors of the company are supposed to be qualified decision-makers, and I trust their competence. As a matter of fact, I believe that, if we were better informed, we might arrive at the same conclusions as they do. There is no question in my mind, however, that we should be better informed. This problem, unfortunately, exists in most large companies.

[Interviewer: Do you think the CEO [Chief Executive Officer] should have a technical background, or on the contrary, be a financial or administrative person?]

The technical end of our industry requires so much knowledge that I do not think that any engineer capable of grasping all of its technical aspects could also have the necessary financial and business background to run the company. Those are two different entities, and both require people with top knowledge; but ultimately, the CEO has to run the company and therefore must be a businessman.

[Interviewer: Do you think TELECO puts itself out for its employees, or does it just meet its formal obligations to them?]

In this matter the company's actions are the direct result of pressure from the employees themselves. If this pressure did not exist, the company would do the minimum required by law. As in any highly technical industry, the company needs highly skilled workers. In order to get them, and keep them, it must therefore treat them adequately, listen to them, and take into account their demands. Most of those demands are made collectively, through union channels, and thus force top management into concrete action.

[Interviewer: And how about you? Do you ever do more for the company than your duties require?]

If I ever have, I did so not for the benefit of the company *per se*, but because of a work overload which sometimes occurs and must be absorbed somehow. On the other hand, there are also times when work is slow, so we are even . . . I don't mean to say that I am getting as much as I am giving. In addition to my salary, I also get satisfaction from what I am doing. I appreciate this opportunity, so all in all, I guess I am getting as much as I am giving. This question involves intangibles and could be discussed *ad infinitum*.

[Interviewer: What about engineers' salaries here, compared to the industry in general?]

As far as my own goes, it may be a little bit below the norm . . . Am I underpaid for the work I do? It depends on the frame of reference. Compared to my colleagues, compared to a maid or a skilled worker? Compared to the skilled worker, I do not think his salary is high enough. Not that I think I am overpaid. I would tend to say the contrary. In absolute terms, if a minimum salary took care of everyone's basic needs, then a scale of higher salaries could be arrived at according to certain criteria. But which criteria? The responsibilities one has, the tediousness of one's work? As far as I am concerned, my work is not tedious, or performed in unpleasant conditions, and it involves certain responsibilities. But how one evaluates and quantifies the responsibilities, I don't know. The system of monetary compensation for job performance has been slowly built up over many years and is now well established.

At TELECO there is little difference in salary between a technician and an entry level engineer. And rightly so, because the engineer is destined for a specific career; he is going to go up the ladder, while the technician's progress will be much slower. Within ten years, a big difference will have emerged. And it is this difference in rates of progress which is not quite fair

. . . As to differences among the engineers themselves, there again I feel the gap is too big. However, if you ask me about the difference in salary between an ordinary engineer and that of the head of a department, I don't know what it amounts to, but whatever it is it's entirely justified because of the responsibilities, the workload, the time spent on the job, which includes weekends.

[Interviewer: In the military, they talk about the morale of the troops. Can one talk about the morale of the *cadres* at TELECO, and if so, what's it like?]

No, I don't think so, because the *cadres* do not form a cohesive entity, a troop. To group all engineers under one collective umbrella is wrong, for they're far too individualistic. However, we can talk about the morale of a division, which includes technicians as well as engineers. There might be slight differences between engineers and technicians, since the former have a better understanding of the project, and a greater involvement in it. The morale in my division is fairly good. It's far from being euphoric, however.

Labor relations are good here. This is due in great part to the fact that the employees do not make great demands. It is obvious that the unions do not put much pressure on the company because of that. If the unions had the backing of the employees, the pressure would be greater, and the relations between the two would be adversely affected. As I have said before, there exists an adequate equilibrium and a mutual understanding between the workers and management at TELECO. As far as where my loyalty or affinity lies, I think it is on the side of the technicians. I feel closer to them than to top management, not because I know them better, and have daily contacts with them, but because I consider myself a salaried person, just as they are, and not as a direct subordinate of top management. And I believe this is the case for at least 80 percent of all engineers. You'd have to go as high as heads of departments or higher to find a greater identification with those who run the company.

[Interviewer: Do *cadres* themselves need to be organized and represented?]

As salaried people, yes, they should be represented, along with all the other salaried people. I don't believe that they should form a separate group, a separate union of engineers or *cadres* to discuss their problems or demands.

[Interviewer: May I ask you who you vote for in professional elections, that is, for which union list?]

I have not been consistent in my votes. I have not voted for the CGC [the *cadres*-specific and relatively conservative Confédération générale des cadres], that's for sure. I voted for the CFDT [the radical Confédération française démocratique du travail] at the beginning when it was formed here. Then I believe that in the last elections I voted for the CGT [a workers' confederation loosely associated with the French Communist Party], because I did not agree with the spirit of what the CFDT was doing here in the context of TELECO [probably a reference to the fact that the CFDT section at TELECO was very militant]. My feelings are that the CGT program here is more constructive than the CFDT's. Their proposals are more constructive; even though their primary role is to protect the employees, still their positions are constructive and reasonable. At least here at TELECO; this may not be the case on the outside or nationally.

[Interviewer: Are you interested in politics? If so, are you active in any way; for example, do you ever attent political meetings or participate in electoral campaigns?]

I am interested, but ever since I started working I've done no more than vote. As a student I was more involved. Let's say I was more of a militant than now. I had some responsibilities, and participated in political gatherings. Obviously my attitude has changed since that time due to circumstances. As soon as you start working, you realize that the positions and opinions you had as a student were not very realistic. At least that's what happened to me. Furthermore, since my job involved a move to the Paris area, I had to face other worries and concerns, and therefore put aside all political and union activities. Ideologically, however, I haven't changed much.

My parents? They were always interested in politics. It was an inherent part of our life, of our discussions. Their convictions were centrist, and as Catholics, they leaned towards the right. I assume they have evolved with the times, but still there must exist certain traditional barriers with rightist tendencies.

[Interviewer: What about you? Which political party do you prefer?]

The Socialist Party. But I am not affiliated with a specific faction within the party. Why the Socialist Party? I guess it is as a result of my political convictions and involvements when I was a student. I see the party as a means to bringing about change and to creating a better, more humane society. I would like to see the obliteration of the present coercive forces that exist between the unions and management. They are the result of a specific environment which must be changed in order to end the destructively hostile relationship between the two."

The interview continues with a few questions about specific issues in French politics and with several questions about family, friends, and life off the job. In that the answers to these are less relevant to the central themes of this book, a brief summary of them will suffice.

Pierre voted for the Socialist Party candidates in the first and second rounds of the latest national assembly elections, and for the Socialist Party leader, François Mitterrand, in the Presidential elections. He is uncertain about his capacity to influence governmental decisions, feels that the big corporations have too much economic and political power, and regards the tax system, with its emphasis on indirect taxes, as unfairly regressive. He approves of the Socialists' proposal for nationalizing the big private companies that have a significant influence on the economy and national affairs, but is against the Communist Party's ideas for nationalizing many more companies. "Then state capitalism would replace private capitalism, and we would be back where we started. I am against all monopolies; favor competition among all – a kind of pluralism." He was also highly skeptical about proposals for workers' control within companies. "It's unrealistic; rules are necessary. It is very desirable to have more participation and consultation than exist now, but someone must make decisions and take responsibility."

Pierre questions the need for many modern "gadgets" (including the Concorde), and favors more conservation of natural resources. Yet, he is not against economic growth so much as concerned about its direction and management. He believes science can solve the problem of radioactive waste disposal, and so favors nuclear power plants.

He does not see his professional life as playing an important role in his private life. When he goes home after work it is not difficult for him to forget about his job. He occasionally speaks of his work or career at home, but not often. He feels he needs to have an interesting job, but any sacrifices he made would have to be in his professional life and not in his private life.

Pierre is ambivalent about his wife's full-time job as a technician. "The pro side is that it gives her a feeling of self-realization, of fulfillment . . . On the other hand, her working creates an added burden to both of us as far as housework is concerned, and I don't think this is the best solution for children." As for his child, "at his age, it's only really his health I worry about. Later on he'll have to learn that life is not easy, that he will get rewards through work, even if it is not easy and presents problems at times. An easy life leads nowhere."

In his leisure time, Pierre enjoys going to the cinema or theater, playing tennis with some of his colleagues from work, and going out with his friends. He also does odd jobs around the house, not only to save money, but also for the sense of satisfaction he gets from completing a project. He reads a sports magazine regularly and also the *Nouvel Observateur*, a weekly current events review that is sympathetic to the Socialist Party.

His best friends – all engineers or technicians – are men he met at engineering school or knows through work. There are three couples that know each other well and do things together on weekends. His friends from work are all from his team, and they and their wives all know each other well. He has little contact with his neighbors. He has little chance to meet them, but when he does it's simply "hello." His wife has more chance to meet them, but she is not really friends with any. He does not belong to any clubs or associations, except for the informal circle of friends from work who play sports together.

On the subject of social classes, Pierre thinks that there do exist "groupings of people who share the same interests." He distinguishes two in particular: "those who have the power to decide and those who don't." He sees himself and even his department manager as belonging to the latter.

The last question is about what activities in his life give him the most satisfaction. Pierre replies that from time to time his work is a source of enormous satisfaction, but that more often his happiest moments are in good discussions with friends.

Many years later and just before this book went to press, I was fortunate enough to be able to meet with Pierre again. He treated me to a superb lunch in a charming little restaurant a few miles from his office, and told me something of the changes in his work life since the original interview.

Pierre is now the *chef de service* of a sales department in a different subsidiary of the conglomerate that owned TELECO. TELECO no longer exists, having been broken up in 1984–85, much of it being sold to a major competitor in the telephone industry. According to Pierre, this reorganization meant lay-offs for many workers, but like most of the engineers, he saw it coming and was able to find his new job before the old one ended. This reorganization of TELECO was just one in a series of purchases and mergers that occurred during and after the parent company's brief spell as a nationalized firm (1982–84). According to Pierre, nationalization did not change the work lives of engineers and *cadres* in any way.

Although the imaging equipment his department sells is highly sophisticated, Pierre's new job is less technical and more people-oriented than his former one. He likes this job very much, and says that his only regret is that he did not change earlier. With respect to his position on the labor market, however, there remain some uncertainties. Just a few days before we met, a major American producer of electrical equipment bought the division in which Pierre works.

Pierre and I did not discuss politics over lunch, but I came away with the distinct impression of a moderate who has grown more concerned with achieving company competitiveness through efficiency and effective management, and less concerned about social justice and reform. Cleaner shaved, better dressed, and more socially skilled than the engineer I had interviewed almost a decade earlier, this prosperous, middle-aged manager appeared quite content with the world and his place in it.

Although Pierre is not typical of anyone, his views are more representative of engineers at TELECO than at PAMPCO. To learn how and why, it is necessary to analyze the larger samples. Chapter 4 begins that process by examining the knowledge these engineers acquire at school and use in their work. That and subsequent chapters refer back to Pierre's comments from time to time.

4

The social meaning of technical knowledge

Several of the theories discussed in chapter 1 link the rise of high technology industry to the *emergence* of a new, knowledge-based society. By this, they do not simply mean an advanced industrial society in which formal education, technical workers, and scientific research play a greater role in the economy. Rather they mean a new society in which knowledge replaces property as the major determinant of social position and values.

The fullest and most influential analysis of "post-industrial" society as 'knowledge-based" is Daniel Bell's (1973). What gives particular force to Bell's vision is his emphasis on the growing centrality of *theoretical* knowledge. Bell argues that the "high technology" industries rely on advances in scientific theory much more than do such empirically developed industries as steel and automobiles. The growing primacy of theory over empiricism and the resulting centrality of research and universities gives rise to a new and more cosmopolitan "technical intelligentsia." The primacy of theory also explains the stratification internal to the technical intelligentsia: an "elite" of research scientists, a "middle class" of engineers, and a "proletariat" of technicians (Bell, 1973: 213). Finally, the centrality of theoretical knowledge accounts for the newly intimate relationship between science and technology, and a broadened commitment among technologists to "functional rationality and technocratic modes of operation" (Bell, 1973: 214).

This argument implies that engineers in science-based industries use more theoretical knowledge in their work than do engineers in more traditional industries. This chapter examines some empirical evidence, looking at both the relative proportions of *ingénieurs diplomés* and *ingénieurs autodidactes* in low and high technology industries and the statements these engineers make about the kinds of knowledge they use and need. First, however, it briefly considers the evolution of French technical education to ascertain the extent and role of theoretical work in the training of engineers.

Technical education in France

In striking contrast to British practice, a formal education in mathematics and general science has long characterized the training of French engineers. Indeed, a common criticism of French engineers in the nineteenth century – one voiced by both industrialists of the time and more recently by economic historians attempting to explain France's slow industrialization has been that their training was *too theoretical* for application in industry.

Theoretical training in the elite schools

This traditional emphasis on theory is clearest in the case of France's elite, state engineering schools, the Ecole Polytechnique and its sister institutions, the Ecole des Mines and the Ecole des Ponts et Chaussées. As early as the 1790s, the Ecole Polytechnique was requiring two years of course work in each of the following subjects: plane geometry, algebra, calculus, chemistry, and mechanics (Shinn, 1980a: 16). Mathematics was emphasized at the expense of the more empirical sciences, but, according to Shinn, the *polytechniciens* took pride in the superiority ("universal truth") of abstract and deductive knowledge to the more pragmatic and inductive knowledge of medicine and law. That graduates of the elite state engineering schools were stronger in theoretical than practical knowledge was of little direct consequence for French industry, as most of these engineers disdained work in private industry in favor of careers as technical and administrative officers in the military and public services. However, the indirect effects were powerful. According to John Weiss (1982: 14), the Polytechnique "became the single most important source of the image of the French engineer as a scientist, formally educated in his *science* by long, intensive years of study."[1]

The first effort to provide France with a stream of highly qualified *industrial* engineers was made by the scientists and industrialists who, in 1829, founded the now elite Ecole Centrale des Arts et Manufactures. Critical of the Polytechnique's concentration on subjects of little use for industrial research and development, they introduced into the Centrale's curriculum a broader array of courses in physics and chemistry, courses intended to provide instruction in both theory *and* application. However, as Weiss (1982: 90) explains,

Despite the "Arts et Manufactures" in the name, the founders clearly did not intend to confine themselves to transmitting the customary methods and private formulas of the industrial trades as they were then constituted, nor even to a systematic selection of current "best practice" technology. In their view, the Ecole Centrale would break completely with such artisanal traditions and with the apprenticeship system of education – which incidentally was still producing almost all British engineers.

The educational program that the Ecole Centrale developed reflected its founders' aspirations to produce an industrial engineer whose claims to social status would rival those of the elite state engineers. The solution was *la science industrielle*, a comprehensive and unified science specific to engineering. "The rejection of the primacy of specialization in engineering education thus became one of the cornerstones of the Ecole Centrale, and consequently, of the entire French civil [as opposed to state] engineering profession" (Weiss, 1982: 97).

The generalists produced by the Centrale were by training and inclination ill suited to meet French industry's need for practicing metallurgical, mechanical, electrical, and chemical engineers. Yet, as the nineteenth century unfolded, the Centrale's curriculum became more, not less theoretical. The emphasis on *la science industrielle* was downgraded in favor of more mathematics and other "impractical" (and therefore less vulgar) subjects, that served less as a core of professional knowledge than as credentials for leadership (Shinn, 1980a).

This association of theoretical knowledge with elite status is not unique to France, but it has been particularly strong there. One reason would seem to be the powerfully persisting influence of aristocratic ideals and social organization during the formative stages of the industrial revolution: the periods of the Restoration (1815–48) and the Second Empire (1852–70). As Alexis de Tocqueville (1945: 2, 45–47) observed in his famous comparison of French and American society in the 1830s, the aristocratic is associated with the theoretical, the democratic with the practical. To be sure, the perception of theoretical knowledge as superior to practical knowledge is often associated with a complete divorce between science and industrial engineering, as in Britain. This was not the case in France, however, perhaps because the elite was so tied to the powerful French state, and the state's involvement in frequent land wars led it to give a high priority to military engineering. Kindleberger (1978: 235) captures the Anglo-French difference nicely when he writes: "The point of contrast is that scientific and technical education were approved *for* the elite in France, *by* the elite in Britain."

In contrast to Britain, there developed in France a technocratic ideology that viewed scientists as the proper elite of an emerging and more advanced order, industrial society, and engineers as the respected practitioners who translated scientific theory into industrial practice. As Weiss notes (1982: 94), "Perhaps the most influential pronouncement on the question of the engineer's scientific identity came from Auguste Comte, who later sent to the Ecole Centrale many of the students he met as an admissions examiner at Ecole Polytechnique from 1836 to 1844." Long before the rise of what Bell calls science-based industries, Comte (Weiss, 1982: 95) wrote:

Between scientists properly so called and the effective directors of production enterprises an intermediate class has begun to form itself in our day, that of engineers,
whose special mission it is to organize the relations between theory and practice. Not
at all concerned with the progress of scientific knowledge, they consider it only in its
current state of development in order to derive from it the industrial applications that
it is capable of providing.

In reality, the theoretically trained *centraliens* left the practical engineering to others, and concentrated on industrial management, but the theoretical claims of *la science industrielle* helped to legitimate their superior
position, a position owed largely to their higher social origins. In short, such
aspects of Bell's vision as the centrality of theoretical knowledge and the
growing power of the technologists were manifest in both the ideology and
the organization of French engineering in the mid-nineteenth century. They
do not represent a response to new technologies, but rather a distinctively
French definition of the role of the engineer.

The *gadzarts*

The bulk of the practicing engineers in French industry during the nineteenth century came from technical high schools, especially the Ecoles
d'Arts et Métiers. Although of modest social origins and usually beginning
their careers as skilled manual workers or draftsmen, most *gadzarts* ("gas
[guys] des arts") rapidly rose to the rank of *ingénieur*. As Day (1978: 459)
explains, "the *gadzart* alone among French engineers was practically as well
as theoretically trained, and for this reason was much sought after by the
mechanical industries."

Unfortunately for the *gadzarts* of the nineteenth century, this training was
increasingly too elementary to earn them the respect of the *ingénieurs
diplomés* from the Polytechnique and the Centrale or to satisfy the demands
of French industrialists. To fill the gap between such inadequately educated
technicians and the overly theoretical and ambitious *centraliens*, local
French governments and Chambers of Commerce established many new
engineering schools between 1880 and 1914. These new schools offered
eighteen months of intensive instruction and laboratory work in basic
physics, chemistry, mechanics and mathematics, followed by an equal
period of more specialized training in a specific branch of engineering,
including a three to six month apprenticeship in an industrial firm (Shinn,
1980a: 22). The graduates were *ingénieurs diplomés*, "none worked as
skilled workers," and the vast majority "filled middle level and upper level
posts within firms" (Shinn, 1980a: 23).

In response to this challenge, the well-organized alumni of the several
Ecoles d'Arts et Métiers succeeded in getting those schools upgraded to

official engineering schools, authorized to award a diploma in mechanical engineering after 1907. After World War II, the post-Liberation government further upgraded these schools, raising the entrance requirements, extending the program from three to four years, and reorganizing the six separate schools into regional campuses of a single institution, renamed in 1963 the Ecole Nationale Supérieur des Arts et Métiers.

Has the training of the *gadzarts* become more theoretical? According to Janine Herbay (Laffitte, 1973: 64), instruction at the Ecole Nationale Supérieur des Arts et Métiers is divided into two basic areas, general science education and more practical "technical training" in workshops and laboratories. The science courses – applied math, physics, chemistry, general mechanics, applied mechanics, metallurgy and electrical engineering – appear to emphasize applications. Herbay, herself, stresses the practical virtues of the Arts et Métiers engineer: the *gadzart* is not a scientist of mechanics, but rather "*un ingénieur généraliste à dominante mécanique*" as well as an "*ingénieur constructeur*," *par excellence*, a "*réalisateur*" (a "doer" or producer), and a communicator who can (and willingly does) explain to workmen the technical complexities of materials and mechanisms. These are not the virtues of a cosmopolitan technical intelligentsia. Yet, the Arts et Métiers is today the single most important engineering school in France, educating about one in six of the *ingénieurs diplomés* produced by that country's 154 engineering schools (Laffitte, 1973: 63).

If today's *gadzarts* are better trained in basic science than they were half a century ago, it is largely because the school's campaign to upgrade its status has required more emphasis on math and physics in a country where theoretical knowledge remains prestigious. What has traditionally made the *écoles* in France more respectable than the *Université* is the requirement of passing a highly competitive entrance examination – the *concours*. Students normally prepare for the *concours* by taking two or even three years of special courses after the *Bac*. It is in these *cours préparatoires* that French *ingénieurs diplomés* get much of their high level training in mathematics and theoretical physics. In 1974, the Arts et Métiers was admitted to the inner circle of *grandes écoles* that require two years of *cours préparatoires*.

While the Arts et Métiers has been raising its status, other, younger, schools have sprung up to train the engineers of modest origins and ambitions. As Fritz Ringer (1979: 126) notes, "their ancestors are not the Ecole Polytechnique and the Ecole Centrale, but the *écoles d'arts et métiers* . . . These institutions have not normally been considered *grandes écoles*, except in the most inclusive sense of that term and since the Second World War."[2] Indeed, they are often referred to as *petites écoles d'ingénieurs*, although a few (e.g. the Ecole Supérieure d'Electricité) have become elite schools. What make them *petites* are their relatively low admission standards, and curricula oriented towards practical engineering. Many of these schools are called "specialized" by the French, but what they mean is not that these schools produce specialists in a particular subdiscipline, such as electrical engineering, but rather that they produce *polyvalent* engineers with a practical orientation towards a specific sector or industry. Much of the post-war expansion in France's production of *ingénieurs diplomés* has been due to the expansion of these schools.

Table 4.1 shows that between 1958 and 1978, enrollments in the *petites écoles* (column C) increased six times as fast as in the elite schools, and now account for over two-thirds of all enrollments in French engineering schools.

Many French engineers do pursue additional training beyond that represented by the *diplôme d'ingénieur*, often while in engineering school, and as many as 36 percent of the *ingénieurs diplomés* have obtained a second degree, most commonly in a scientific discipline (FASFID, 1977: 28–29). The growth in these second degrees may indicate a growing demand for scientifically sophisticated engineers, or it may simply represent an extension of credentials competition, but the overall impression remains one of an educational system in which more engineers are being trained in practical subjects even though theoretical knowledge is still accorded greater prestige.

Training for research

Although the great growth among French engineering schools has been among those giving relatively little emphasis to theory, research and training for it have become more important.

Established in 1957 (Lyon), 1963 (Toulouse), and 1966 (Rennes), the Instituts Nationaux des Sciences Appliquées (INSA) are unspecialized engineering schools designed specifically to furnish modern French industry with trained research engineers. Two points about them are relevant to the debate about the growing centrality of theoretical knowledge. First, although large by French standards, these schools – the most research-oriented of all the *écoles d'ingénieur* – produce fewer than 10 percent of French *ingénieurs diplomés*. Second, they are *petites écoles*. In the rankings reproduced in Table 7.2 (p. 142), the INSA stand next to last in the large middle group of schools; in PAMPCO's rankings, they are classed in the third (i.e. bottom) group. One reason is that the INSA are among the few French engineering schools that admit students on the basis of their previous academic record and a personal interview. True (Lafitte, 1973: 81–84), the five-year program at the INSA includes a two-year *cycle préparatoire* that emphasizes math and general science, but it is less rigorous and theoretical than that offered by most of the *cours préparatoires*. Yet, it is from the ranks of these INSA students that high technology industry recruits many engineers for work in its research and development laboratories. In other words, while the science-based industries have given rise to an increasing

Table 4.1. *Enrollments of engineering students by school status*

	Ecoles d'ingénieurs			Facultés des sciences	Ecoles de commerce
	A	B	C		
1957–58	3,285	4,840	8,301	61,725[a]	4,907
1962–63	3,683	6,250	16,338	88,595	5,538
1967–68	4,204	6,883	18,116	137,111	9,952
1972–73	4,798	7,341	23,065	120,142	9,542
1977–78	4,893	8,165	25,478	126,072	10,990

[a] 1958–59
Source: Adapted from Boltanski, *Les Cadres* (1982: 318).

number of research engineers, the result in this case is neither a growing centrality of theory nor a more cosmopolitan and prestigious technical intelligentsia.

To be sure, the elite schools continue to offer highly advanced training, including heavy doses of basic science. About one-third of the course work at the Ecole Centrale today involves instruction in basic science (Laffitte, 1973: 80). Moreover, since World War II, the Centrale has developed several of its own research institutes and groups, and has increasingly provided a portion of its students with specific training for careers as *ingénieurs de recherche*. Some go on to complete advanced degrees in engineering.

Nevertheless, neither research nor graduate work plays as large a role in French engineering schools as in their American counterparts. In a recent article, Petitjean (1986: 54) quotes a French official who is worried about France remaining economically competitive in a high technology world "because our engineers do so little research . . . Search throughout France and you will find fewer [engineering] students in research courses than at MIT alone." Yet, to judge by French prowess in telecommunications, aeronautics, and nuclear power, France appears to be doing quite well in the high technology arena.

As Leon Trilling (1979: 233) observes, "The American pattern in which higher education is joined to research and to training for research in graduate schools is an adaptation of the German model rather than of the French model, which was adopted only at West Point." Indeed, the French model of industrial enterprises and of schools for training their officers is a military model (Grelon, 1982: 720–721). More importantly, most French engineering schools remain highly independent of – and often physically isolated from – the French universities. This bifurcation within higher education is unique to France. At the same time, French engineering schools maintain close ties with the sectors of public administration or private indus-try in which their graduates are concentrated. The latter relationship reinforces their tendency to produce (and thus attract) sector-specific and management-oriented generalists. "In this situation, research is not an essential concern . . . " (Trilling, 1979: 235). The research that is done tends to be very practical, often funded by contracts from a sponsoring company or government office.

This is not to deny the theoretical character of many of the science courses in the elite *grandes écoles*. After comparing the curricula of MIT and France's prestigious Ecole Supérieure d'Aéronautique, Trilling, who teaches at MIT, concludes that "in technical content the differences between the French and American programs are minimal" (1979: 238). It is to suggest, rather, that the meaning of the theoretical knowledge imparted to elite French and American engineers is quite different. For graduates of the elite French schools, theoretical training remains a technique for legit-imating the fast-track careers of future executives. True, some of them will work for five or ten years in research, before being promoted to more managerial positions. But this very career pattern means that even the small minority of theoretically trained research engineers in France develop not the cosmopolitan views and bonds of a "technical intelligentsia," but rather the organizational and entrepreneurial orientations of future managers.

Techniciens supérieurs and *doctorats*

The above discussion noted the disproportionately rapid growth of the *petites écoles* in recent decades – precisely those engineering schools that give the least attention to theoretical science. This pattern of expansion is even more evident if we look at the two extremes of technical education in France.

There has been a vast expansion in the production of *techniciens supérieurs diplomés*, high level technicians who have completed two years of post-secondary technical training at either a *lycée technique* or one of the newer *Institut universitaire de technologie* (IUT) programs emphasizing applied rather than theoretical work. Of the latter, the first was esitablished in 1966, but by 1978 there were 72 IUTs producing more than 20,000 graduates (DUTs: *Diplomés d'un IUT*) per year, twice the figure for *ingénieurs diplomés* (Halls, 1976: 158). At the same time, the *technical lycées* were rapidly increasing their annual production of technicians possessed of a BTS: *Brevet de Technicien Supérieur*, from about 12,000 in 1966 to over 20,000 in 1975 (Boudon, 1977: 110–111). Thus, while the production of *ingénieurs diplomés* increased from 6,200 per year in 1961 to 10,000 in 1975, or 61 percent, the total production of *techniciens supérieurs diplomés* increased from about 4,000 in 1961 to about 40,000 in 1975 or 900 percent (Boudon, 1977: 110).[3] Not all those earning a BTS or DUT became industrial *techniciens supérieurs* by any means, for the IUTs in particular offered schooling in a broad range of "technical" specialties, but it is nonetheless likely that the rate of growth for these *techniciens supérieurs* outstripped that for *ingénieurs diplomés*, although probably not that for graduates of the *petites écoles d'ingénieur*. In short, the recent growth in French technical education has been most pronounced in schools that offer training in fields of applied rather than theoretical knowledge.

Finally, it is worth noting that the production of engineers with higher degrees, i.e. *docteurs ingénieurs*, remains at a very low level – 466 in 1977 (FASFID, 1977: 13). True, many engineers have obtained second, if not higher, degrees, but more significant are the figures on *cours de perfectionnement*, continuing education courses taken by employed engineers at company expense and often during work time. Fifty-six percent of the respondents to a major survey of *ingénieurs diplomés* said that they had taken such a course at some time during the last three years. Of these, 124 percent (multiple responses) mentioned courses in management(40 percent), languages (29 percent), human relations (25 percent), economics (21 percent) and/or marketing (9 percent), while only 62 percent mentioned courses in science (24 percent), technology (17 percent), and/or computer programming (21 percent) (FASFID, 1977: 31). In other words, practicing French engineers reveal relatively little concern for raising their level of training in the theoretical areas of professional knowledge.

Taken together, all this material about the formal schooling of French engineers suggests serious problems with Bell's argument about the growing centrality of theoretical knowledge. While it is true that French engineering students, especially those at the more prestigious *grandes écoles*, do take a substantial number of courses in mathematics and basic science, it is also clear that such theoretical training has a long and honorable history in France, and that the vast expansion of technical education since World War II has been in the less theoretical schools and IUTs. In short, the post-war rise of the science-based industries is associated with a decline in the emphasis upon theoretical science in the training of most French engineers.

The knowledge that PAMPCO and TELECO engineers use

However great the decline in theoretical studies, it is nevertheless likely that by comparison with their British counterparts, French engineers receive a good deal more training in mathematics and basic science as well as more formal schooling. Yet, training does not imply use on the job of the knowledge learned. How much of the theoretical or even formally-learned knowledge do French engineers use in their work, and do those employed in the high technology sector rely more on scientific training than those in the traditional sector? Data gathered through the field research and interviews at PAMPCO and TELECO bear directly on these questions.

All but three of the engineers at PAMPCO and TELECO said that the knowledge they used most often in their daily work was that gained through experience at work. Of the three exceptions – two at TELECO and one at PAMPCO – only one, a *technicien supérieur* at TELECO, emphasized the importance of the theoretical knowledge he had acquired as a result of formal training; the other two gave priority to formal training as opposed to experience, but stressed the value of the technical skills and codified knowledge thus learned rather than the theoretical knowledge.

Responses to a written question reveal a less dramatic but similar emphasis at both firms upon the importance of experience. Asked whether the technical work they do could be done by someone with less technical education than theirs, only 38 percent replied in the negative; 40 percent felt that experience was an adequate substitute, while most of the remainder replied that their work could be done by someone with less technical knowledge. Technical education does not necessarily imply theoretical training, referring as it does to schooling in applied science and technology as well, but learning by way of experience does seem to preclude theoretical knowledge.

Interestingly, it was engineers at the traditional firm, PAMPCO, who were most likely to regard their level of formal education as indispensable. Only 39 percent (seven of eighteen) of PAMPCO's *ingénieurs diplomés* agreed that experience was sufficient. By contrast, 75 percent of TELECO's *ingénieurs diplomés* (excluding, as usual, *chefs de service* and sales engineers) said experience was an adequate substitute. The emphasis on experience varied considerably by hierarchical category, much less by technical function. Ironically, *chefs* and *ingénieurs autodidactes* emphasized the importance of formal training more than did *ingénieurs diplomés* and *techniciens*.

With respect to Bell's argument, these findings are important and bear elaboration. The following dialogue and comments were fairly typical of the responses given by engineers in the pipe and metal products firm to the question: "(In doing your current job) do you most often use knowledge learned at school, knowledge obtained through on-the-job training programs (*stages*), or rather knowledge gained through experience at work?"

I think I'd put the last in first place, the second in second place, but as to what I learned at school, well, that's why I think an *ingénieur nondiplomé* could do my job perfectly well . . . In our work you never need to refer to higher math or general mechanics theory . . . You can get along fine with a good knowledge of industrial electrics and mechanics. [an *ingénieur diplomé* in production]

Not school. Some practical training programs since then a bit. But most of all, experience. What's important is a natural ability to analyze and create, which some *techniciens* have as well as *ingénieurs*, but also an ability to draw practical conclusions and achieve a synthesis. *Ingénieurs* seem much better than *techniciens* at the latter, although I'm not sure whether the reason is their training or natural talent. [an *ingénieur diplomé* in design]

That gained from experience, but the capacity (*l'esprit*) for analysis, for synthesis, for reasoning, that comes from schooling. [an *ingénieur diplomé* in design]

These answers are revealing about the mystique that surrounds higher technical education in France and its function as a device for legitimating status differences. Many PAMPCO engineers reported that it was from their math and science courses that they had learned a certain *esprit de raisonnement*, a logical approach to problem-solving. They did not remember much calculus and they thought most of the technical skills they now used had been learned on the job; nevertheless, like many American lawyers, they insisted on the value of their professional education – one they regarded as general rather than specialized in any one engineering discipline – as learning to think clearly, see the whole picture, and relate part to whole. *Ingénieurs diplomés* in positions of authority (*chefs* and production engineers) were particularly likely to stress the general usefulness of their math and science training in *grandes écoles*. They thought that a technician might know more about a certain machine or process, but this very specialization incapacitated him for the broader managerial responsibilities of *cadres*. By contrast, their own general but rigorous training in math and a variety of sciences uniquely qualified them for the real work of the engineer, management of industry's central functions, development and production.

Even more surprising than the emphasis of PAMPCO engineers on formal training is the emphasis on experience at TELECO. With respect to Bell's argument about theoretical knowledge, this is of particular significance. The following quotes are all from engineers at TELECO, in reply to questions about the sources of knowledge used on the job. The first pair are from *ingénieurs diplomés* who work in research at this advanced electronics firm.

Now that I'm in advanced research, I use and need most all the knowledge I have acquired in a variety of positions at TELECO. I like that, but I regret not having more opportunity to use the theory I learned at school. What's needed in this research is a logical mind, not a theoretical base, even for learning about a new area. We have some contact with scientists at international conferences and such, but we learn only the results of their work and then modify our systems accordingly. What's most useful for me is my broad experience with the whole range of telephone needs and problems, experience gained working with four others to adapt (a less advanced) telephone switching center for use in Martinique.

I use little of what I learned in school, especially in any precise way, especially math. Little physics either. You don't need to know how a transistor works in the telephone industry. Yet, science training helps. I work at a general conceptual level where one needs little specific knowledge of scientific laws, but where systematic scientific thinking helps.

Of particular interest is the *chef* of the closest thing to a pure research group at TELECO, the long-range development department. Significantly, in a product division of over 300 *ingénieurs*, his team consisted of only ten *ingénieurs* (all *diplomés*). The *chef* of this department was more insistent than most on the need for some theoretical knowledge. Yet, even he emphasized the value of practical experience with telephones. His own technical education had been at a *petite école* known for its focus on practical engineering rather than science and management. Saying "we don't seek to do university research here," he stressed his preference for practical young engineers with first-hand experience of working with telephone exchanges.

The rest worked in development, design, refinement, methods, production, maintenance, testing, and service. A brief look at the responses of some of the *ingénieurs diplomés* in development is instructive.

My training in basic sciences? That I've never used. Well, the math, yes, because at one point I worked in the area of calculating traffic flows. But apart from that, all the courses in theoretical physics that I took, even all the courses in basic electronics, well, they help to understand the phenomena, to interpret them, but I think the main use has been to train my mind, as they say.

Almost entirely experience, memory and good sense. Maybe it could be taught in a school, but it really helps to live and work in it . . . It was fun learning science, but you can't apply it.

Experience. My work is conceptual, but it involves little math or physics. I have to learn how to use the available circuits, including how they work, but not really why. Ecole Telecom (an elite *grande école*) gave us a general education, taught us how to reason well, not specific knowledge.

The final few quotes are taken from replies by development-design engineers at the Simulator Division. The first is from the head of the department responsible for *études électroniques*, an engineer who had published a technical book and who teaches two weeks a year at Sup'Elec, an elite *grande école*.

It's hard to say. There's something we learned at the *grande école* that is not very tangible – a method of reasoning. When one first leaves school one wonders what one learned there, for one has to learn so much in industry. But the method of reasoning is what distinguishes *ingénieurs* from *ingénieurs maison*.

Experience. The high tech (*de pointe*) part of the work is pretty small. There's some development here, but nothing revolutionary, except at the level of application and utilization.

Experience . . . Even after only three years out of school . . . There's nothing to discover in our work. We study a system that is already advanced. It's a problem of translation. I feel a little frustrated, but I accept that schooling was to prove oneself and to learn basic math and a way of thinking.

These comments are all from *ingénieurs diplomés* working in the development or design of very high technology products. Those working in areas closer to production and after-sale service emphasized even more the value of practical over theoretical knowledge. One *ingénieur diplomé* who installed and tested simulators remarked that such work as his is "still quite unsophisticated." Yet, even those in development and design stressed practical knowledge acquired through concrete experience, and minimized the value of their education in basic science, except as good training in the cognitive processes of analysis and synthesis. In this, they closely resembled their counterparts at the traditional firm.

It might be thought that research, where at least the *chef* regarded some theoretical knowledge as necessary, represents the most rapidly growing function within engineering, but this is not the case. The recent rise of science-based industries has not changed the ratio of research engineers to all engineers. According to the FASFID surveys (1968: 41; 1971: 34; 1974: 70; 1977: 68; 1980: 88), the percentages of *ingénieurs diplomés* working in research and development in 1968 and 1980 were the same, 10 percent. If the base of calculation is confined to *ingénieurs diplomés* working in technical functions (i.e. practicing engineers), the percentages in research and development rise, but the resulting figures are again the same for 1968 and 1980: 17 percent. The 1980 figure for the electrical-electronic industries alone is only 9 percent, compared to 12 percent for mining and metallurgy. It is true that there has been a decline in the percentage of *ingénieurs diplomés* working in production, maintenance, and related activities – from 40 to 33 percent – but the compensating rise has been in the proportions working in design, not research – up from 24 percent in 1968 to 30 percent in 1980. The 1980 percentage working in design within the electrical-electronic industries is 33 percent.[4] To be sure, much of their design work involves research of a sort, but evidently not of the kind that requires much theoretical knowledge.

From informal observations and discussions as well as comments made during the interviews, some likely explanations of the emphasis on experience emerge. The "research" in R&D is extremely applied research. Even for R&D engineers in telecommunications, most product innovations represent new applications of materials or techniques discovered and usually developed elsewhere. The great advances in electronic telephones have relied on breakthroughs in the semiconductor industry. Only the largest telephone companies support their own basic research. Surely there is a closer link between science and technology today than in the nineteenth century, but this affects the knowledge base of a very limited number of industrial technologists. This may be because theoretical knowledge is "codified" into practical procedures before being applied by engineers. In any case, the argument presented here does not challenge the claim that industrial innovations have become increasingly reliant on advances in basic science. Rather, it questions the sociological significance of these developments.

This analysis has relied largely on self-reporting in trying to show that the knowledge used by high technology engineers is far from theoretical. However, other evidence seems to support this conclusion. Very few engineers in either firm appeared to make use of scientific journals or theoretical texts in solving technical problems. Rather, as with Pierre (chapter 3), they tended to refer to technical manuals (often provided by the supplier of the components they were using), consult with colleagues, or call the client who had contracted for a prototype or cost estimate. Only the little band of long-range researchers at TELECO reported having contact with major research labs (in this case the parent company's research center in Paris), or drawing on the scientific literature. And even the *chef* of this group prefers engineers who, like himself, have had broad "hands-on" experience with the product and its now obsolete predecessors. Finally, there is the evidence on the place of *autodidactes* among French engineers.

Ingénieurs autodidactes

If the knowledge that engineers use were becoming increasingly theoretical, it would also be increasingly unsuitable for on-the-job learning, for acquisition through experience rather than formal learning. That would mean a decline in the proportion of French engineers who are *autodidactes*, and that the latter should be concentrated in the traditional rather than science-based industries, and in production rather than research.

While evidence on these points is not available for all of French industry, it is available for that substantial portion of it represented by the *Union des industries métallurgiques et minières* (UIMM), an association of employers that covers the aeronautical and electronic industries as well as automobiles, steel, and mining. Between 1956 and 1977 the UIMM conducted and published four surveys of its *ingénieurs* and *cadres*, and the 1977 publication reports several trends during those two decades that bear on the research questions posed above.

First, among all *ingénieurs* in the metal industries, the proportion of those not *ingénieurs diplomés* has grown from 33 percent in 1956 to 36 percent in 1975.[5] One third of these – up from 29 percent in 1956 – have some sort of lower technical degree (up to and including the diplomas of *techniciens supérieurs*), but most are still relatively unschooled former workers who learned their technical skills on the job (UIMM, 1977: 59, 61). Judging from recent educational trends, a growing proportion will have completed secondary education in the future, but this development appears to reflect credentials inflation rather than the needs of industry (Collins, 1979).

Second, a greater proportion of these *ingénieurs autodidactes* worked in research or design in 1975 than in 1956, 30.1 percent as compared to 25.8 percent. This is largely because the fraction of all *ingénieurs* employed in research or design increased from 36 percent in 1956 to 39 percent in 1975; as a proportion of all *ingénieurs* working in research or design, *ingénieurs autodidactes* barely rose, if at all – from 27.5 percent to 28.1 percent (UIMM, 1977: 69). Nevertheless, while *ingénieurs autodidactes* are slightly under-represented in research and design (28.1 percent vs. 36 percent, for all technical functions combined), they are neither concentrated in production nor losing ground in the growing research and design functions.

Finally, what about the distribution of *ingénieurs autodidactes* across industries; are they concentrated in the traditional rather than science-based sector? Here the UIMM report is less helpful, giving data on just five broad industries, and only for all *ingénieurs* and *cadres non-diplomés* together. The percentages of all *ingénieurs* and *cadres* that were *non-diplomés* in 1975 are: steel, 28 percent; metal-working/mechanical, 47 percent; automobiles, 47 percent; electrical (including electronic), 29.5 percent; aeronautical, 41 percent (UIMM, 1977: 77-95). It should be borne in mind that the proportion of all French *ingénieurs* and *cadres* that was non-diplomés was 40 percent in 1975. In view of the confounding of technical and non-technical *autodidactes*, these industry-specific figures do not warrant detailed interpretation. What they do indicate, however, is that *autodidactes* in general, and probably *ingénieurs autodidactes* as well, are well represented in the science-based industries (electrical-electronic and aeronautical). So also are higher education graduates who lack any scientific or technical training. According to figures cited by Boltanski (1987: 193), "46 percent of the data-processing professionals surveyed held degrees in the liberal arts, political science, or law, or had attended business schools, or were self-taught."

Conclusions

This chapter has put forward two basic arguments. First, contrary to the expectations that flow from claims about the growing centrality of theoretical knowledge, French engineering schools place less emphasis on theory today than they did in the past. For reasons peculiar to French history and culture, engineering schools in the nineteenth century were elite institutions that cultivated theoretical knowledge as a mark of their social superiority.

The decades since have seen some expansion and democratization of engineering education. Many of today's engineering schools, especially the newer *petites écoles* and the schools for *techniciens supérieurs* appear less preoccupied with theoretical work.

The major way in which the *petites écoles* still resemble the elite *grandes écoles* is in their institutional isolation from the university system and thus academic science, and in their close association with industry. This is probably one reason for the small number of engineers who go on to complete doctorates. Although French engineering schools and the *cours préparatoires* for admission to them continue to require more work in mathematics than engineering schools in some other countries, this emphasis is neither new nor growing, and it does not indicate a strong bond between academic science and applied technology. Indeed, the tendency to call engineering "applied science" appears to be peculiar to Britain and the United States, where "technical change is widely attributed to scientific discoveries in both countries, in spite of much evidence to the contrary," and where "the tendency to define engineers as narrow specialists and as 'scientists' of a kind appears to have helped to produce the phenomenon itself" (Glover, 1983: 20, 18).

The second argument concerned the knowledge used by the engineers interviewed at PAMPCO and TELECO. These engineers reported making much greater use of the concrete knowledge they had acquired through experience on the job than of their theoretical training at school. To be sure, many insisted on the value of their education in mathematics and science. However, the reason was not that they used such knowledge in many direct ways at work, but rather that it provided good training in logical thinking. To the extent that theory was seen as general, it was also valued as providing the *polyvalent* base for coordinating the more specialized activities of *techniciens*. Such coordination, however, was emphasized by the PAMPCO *ingénieurs* even more than those in the high technology setting, and seems far removed from the kinds of application imagined by those envisaging a new breed of cosmopolitan technologists. Indeed, the overall similarity in the responses of the engineers in these two very different industrial settings is striking.

Moreover, the continuing importance of *autodidactes* in French engineering makes it difficult to accept the engineers' claims that their training in mathematics and science really teaches them to reason more clearly or coordinate technical specialists more effectively. Daniel Gourisse (Petitjean, 1986: 54), director of both the Centrale and the National Conference of *Grandes Ecoles*, admitted recently that "the requirements of the last year of the [*Bac*] C and of the *classes préparatoires* in math are undoubtedly a little excessive."

To the extent that the training is more theoretical than necessary, the explanation seems to be that it serves as a mark of status within and outside the occupation. As such, it is a distinction around which, to borrow Parkin's term (1979: 44ff.), "exclusionary closure" can be organized. In this case, it justifies the *cadre* barrier between engineers and technicians as one between generalists and specialists, conceptualizers and executors, and thus managers and managed.

If that knowledge has little value beyond its legitimating function, why do the managers of industry favor *ingénieurs diplomés* over graduates of junior engineering and university science programs? Part of the explanation lies in the political reasoning of managers, themselves graduates of *grandes écoles*, recruiting graduates of their own schools. As Perrow (1979: ch. 1) suggests, such recruitment practices are useful for securing loyal allies in intra-organizational struggles. In France, moreover, old school networks are organized and powerful, and certain firms come to be identified with specific *grandes écoles* generation after generation (Kindleberger, 1978).

Yet, it is the preference for graduates of *grandes écoles* in general that is so striking in France. Part of the explanation is that the presence on a firm's staff of graduates of any elite school supports that firm's status and competitive position, and facilitates cooperative relations with other organizations who employ graduates of the same schools. Company brochures in France advertise without embarrassment the schools represented on the company's staff, and firms hire graduates of prestigious institutions "to use the contacts that go along with their degrees" (Rivard et al., 1982: 46).

Part of the explanation lies in what Kanter (1977: ch. 3) calls "homosocial reproduction." Kanter argues that top managers promote people like themselves to managerial positions because the discretionary component of such jobs makes trustworthiness critical, and because people more easily trust their "own kind" to think and act as they would.

The question remains, however, why are "own kind" defined in France in a fashion that excludes university graduates? French personnel managers are quite frank about the reasons they prefer graduates of engineering schools to the graduates of the university science programs. Several patiently explained to me how the two to three years of arduous post-*Bac cours préparatoires* for the highly selective *grande école* entrance exams served to weed out all but the most motivated, self-disciplined and hard-working students. The universities, by contrast, accept anyone with a *Bac*. These managers also explained how the educational programs in the engineering schools were much more structured and demanding than those in the universities, requiring students to produce results frequently, not just in examinations at the end of the year. In short these schools are valued for their capacity to select and develop character traits that are not particular to any occupation, but highly desirable in "trusted workers" (Whalley, 1986) of all types. (It is not coincidental that the students in the *grandes écoles*, especially the elite ones, are more likely to come from upper-class families.) The result is a self-reinforcing belief that *ingénieurs diplomés* and other graduates of *grandes écoles* must be smart, hardworking, disciplined, adaptable, and dependable.

Favoring the graduates of *grandes écoles* does not mean excluding all others from the ranks of engineers. On the contrary, the large group of *ingénieurs autodidactes* makes up a critical part of this system. As is explained in chapter 7, the *ingénieurs diplomés* have managed to limit the expansion of the engineering schools, and France has a lower proportion of graduate engineers than other advanced societies. The way management has filled the gap in the formal production of *ingénieurs* is by creating its own through the promotion of *autodidactes*. Yet employers do not mind resorting to this local source, for by doing so they avoid becoming dependent on those with power on the external labor market, split the "occupation" in two, and create a sub-group of *ingénieurs* who are exceptionally dependent on them.

A general theme begins to emerge. *Autodidactes* lack vocational degrees and communities with which to identify and bargain. *Ingénieurs diplomés* have them by definition, but are valued by employers for their presumed general capacities to work, learn, and adapt, rather than their specific skills. Their technical education is not so much academic or scientific as oriented towards specific sectors of industry; as such, it is *polyvalent* rather than specialized. This means much of the practical knowledge needed at work must be learned on the job. Employers may complain about the uselessness of what students learn at school, but they clearly value the adaptability that employees lacking vocational skills must show. Adaptability is highly desirable in the unstable environments of new and high technology companies, especially in societies where custom and law make it difficult to dismiss employees.

To be sure, there are other ways to achieve an adaptable labor force, as Maurice et al. show in the case of Germany, a country in which vocational education is highly developed and integrated with industry. Yet, for reasons of its own national history, France stands out for its lack of vocational training. According to Maurice et al. (1984: 241), only 30.5 percent of the work force has a vocational diploma, compared to 73 percent in Germany. Even for the graduates of the higher vocational schools in France, the *grandes écoles*, there is a weak connection between education and type of employment. About half of *ingénieurs diplomés* do not work as engineers, and nearly half of practicing *ingénieurs* lack engineering diplomas. Rather than drawing French engineers closer to academic scientists, the emphasis on math and general theory in engineering schools reinforces the separation between formal schooling and job skills by increasing the importance of on-the-job learning. Theoretical knowledge remains important in higher French education, but rather than linking engineers to an elite of scientists in a community of professionals, it qualifies them for management in industrial organizations.

5

The organization and experience of technical work

French engineers were quick to show interest in the ideas of Henry Ford, Frederick Taylor, and other American advocates of the use of "scientific" techniques to rationalize the organization of factory work (Moutet, 1985). Such approaches appealed readily to the technocratic mentality that has marked French engineering since the days of Auguste Comte. They also offered engineers a means of increasing their power as critical experts who alone could determine the "one best way" of organizing production and managing labor. As in the United States, employers showed less enthusiasm, but the production crisis of World War I brought support from the French government for scientific management, and the rationalization movement flourished until the Depression. In the process, French industrial engineers participated vigorously in the simplifying and routinizing of much work previously done by skilled craftsmen.

It would be ironic if engineers themselves were now becoming the victims of efforts to deskill their work. Yet, that is what Marxist theories of proletarianization predict. Braverman emphasizes the inevitability of deskilling under capitalism wherever "massification" makes it practical for management to subdivide and routinize tasks. And although a growing number of critics effectively question Braverman's analysis of manual and clerical workers, there remains considerable debate about the applicability of the deskilling thesis to technical and professional employees.[1]

Chapter 1 discussed that debate, emphasizing the view of proletarianization theorists that engineers face increasing specialization of their labor, codification of their knowledge, under-utilization of their skills, and supervision of their work. This chapter and the next examine the evidence for such expectations that emerges from an investigation of the work that PAMPCO and TELECO engineers do and their evaluations of it. This chapter begins with a brief historical overview of the place of the French engineer in the division of labor. It then examines in some detail the organization of technical labor and the actual work engineers do at PAMPCO and TELECO. There follows an analysis of these engineers' own feelings about their work. The chapter ends by considering some labor market explanations for the patterns found.

French engineers, yesterday and today

If engineers rely less on theoretical knowledge than on knowledge gained through experience on the job, what then distinguishes their skills from those of the machinists, mechanics, draftsmen or technicians who also apply technical knowledge to industrial production? The question is one that many French engineers ask themselves, sometimes a little anxiously, and their own answer is as important as an "objective" one. The answer that many French engineers give is that they are *polyvalent*, or multi-skilled, as well as responsible, and this *polyvalence* enables them to manage others. Skilled craftsmen and technicians, by contrast, are specialists. Thus, the engineers alone know enough about a variety of technical subjects – mechanics, hydraulics, electricity, etc. – to serve as the innovative designers and coordinators of complex industrial projects.

This answer reflects more an ideal image of engineering than current reality; judging by the work they do, many contemporary French engineers are neither generalists nor managers. However, the ideal itself is significant, both as a standard against which the present is judged, and as a key to the distinctive character of engineering in France. Thus this section focuses on the content, the origins, and some of the consequences of this ideal.

The ideal of the engineer as a *polyvalent* technical manager is an ideal of a mental worker, not a craftsman or talented tinkerer. The latter do both mental and manual work, designing a product and then physically creating it. And it is the combination of mental and physical engagement in the complete cycle of creation, from initial concept to finished product, that informs one major ideal – the craft ideal – of meaningful work (Mills, 1951: ch. 10). It might be thought that industrial engineers in particular, having in many countries evolved out of the skilled crafts, feel some regrets about the separation of head and hand, and the lack of opportunity for contact with the products and processes they now only design. Such is not the case in France. Although over half the engineers interviewed described themselves as fairly avid *bricoleurs* (handymen) in their youth, that is as young men they enjoyed building and fixing mechanical or electrical devices, very few expressed regrets about the absence of "hands-on" activity in their current work. The exceptions tended to be *ingénieurs maison* who had once been skilled workers and whose positions as engineers involved many frustrations. Moreover, production engineers are much freer than their American counterparts to handle materials and adjust machinery, if they wish. They are not constrained by the detailed delineation of job duties and jurisdictions that characterizes American industry. On more than one occasion I found a PAMPCO engineer with sleeves rolled up and tools in hand, working on a pipe-spinning machine. Design engineers had fewer occasions for such physical contact, but did not care. Several of them stressed the excitement, tension, and satisfaction of witnessing the initial test of a new product, machine, or process on which they had worked. In these cases, however, the satisfaction stemmed less from manual activity than involvement in the actualization of their designs.

The ideal of the French engineer as a *polyvalent* technical officer rather than a highly skilled craftsman is rooted in the particular history of the engineering occupation in France. Unlike their counterparts in many other countries, French engineers did not emerge out of the mental–manual differentiation of skilled craft work. In the nineteenth century French engineers were not artisan-engineers of the type that flourished in Britain, a country in which craft apprenticeships and the craft tradition remain important among technical workers in the engineering industries (Smith, 1987).

Rather, nineteenth-century French engineers tended to be *polyvalent* technical managers. To be sure, engineer is an ambiguous term, and many *gadzarts* filled less managerial posts, especially early in their careers. But *Monsieur l'ingénieur* was often the all-purpose technical director of a small family firm. Working closely with the owner-manager, he managed production and exercised responsibility for a wide range of technical problems. He designed, built, tested and supervised the on-going use and maintenance of new equipment and processes.[2] The history of PAMPCO provides a nice illustration. In 1859, the major founder-owner of PAMPCO entrusted a 24-year-old *gadzart* with the responsibility for managing the factory. This engineer designed a major new blast furnace, and went on to greatly expand and improve the company's production facilities and techniques.

Of the many reasons that French engineers enjoyed such an impressive combination of technical and managerial responsibilities, two bear mention here. One is that in abolishing the craft guilds and the apprenticeship system, the French Revolution dealt a severe blow to the skilled crafts in France. The other is that the occupation of engineer had already emerged as a state-sponsored elite corps of technical officers in the bureaucratic and rationalistic climate of eighteenth-century France. As explained in the previous chapter, these state engineers obtained their positions by graduating from prestigious, university-level schools that stressed math and theory, not by demonstrating technical skills as clever workmen. And as military officers, they supervised rather than performed manual work.

France might have, but did not, develop vocational schools to produce the needed industrial artisans. Arts et Métiers initially represented one such effort, but it quickly evolved into a high school for broadly trained technicians. It was the end of the nineteenth century before any vocational schools for manual workers were established (Maurice et al., 1977: 220). In the meantime the *centraliens* were adapting the *polytechnicien* model of a state engineer's education and work to private industry. Lacking skilled workers, yet possessed of a supply of school-trained technicians, French industry followed the existing and respected state precedent of giving industrial engineers broad responsibility for organizing and managing the production process. In the words of Maurice et al. (1977: 221), "engineers acquired very early in France a status comparable to that of high level civil servants and independent professionals." The contemporary ideal of the engineer as a manager – as *Monsieur l'ingénieur* – reflects this unique history of the French engineering occupation, and contrasts sharply with the situation in Britain. Glover (1983: 8) reviews several studies that show the British tendency to define engineers as specialists who are unfit to occupy the highest positions.

The ideal of *polyvalence* is related to this managerial tradition. Craftsmen specialize in specific skills; managers are generalists, or at least they were in the small scale and pre-bureaucratic enterprises of the nineteenth century. Moreover, specialization has never been as respectable in France as in the United States (Cole, 1979: 102), and the founders of the Ecole Centrale went to great lengths to formulate a unified discipline of engineering – *la science industrielle*. Zeldin (1982: 410) is correct when he observes that "the tradition of general culture, which abhorred specialization, has declined," but it has by no means died, and FASFID still distinguishes between engineering schools that provide a *formation* (training) *générale* and those providing a *formation technique spécialisée*. And it is in the former category that are found the most prestigious *grandes écoles* including the Polytechnique, Centrale, and Arts et Métiers.

However *polyvalent* the training of French engineers, their jobs in industry appear to have grown increasingly specialized. This is not a new phenomenon. The rise of the big electrical, chemical, and automobile firms during the "second industrialization" meant large engineering staffs and a certain division of labor among engineers. But the differentiation has continued. In the automobile industry, for example, "until the beginning of the 1940s, the large majority of engineers were divided between those who worked in design (bureau d'études) and those responsible for production. Now there are six categories" (Fridenson, 1985: 438).

Thus, by nineteenth-century standards, most engineers today are specialists. There are still many small manufacturing firms in France, and in them are to be found most of the surviving examples of *Monsieur l'ingénieur*. But 65 percent of contemporary French engineers work in companies of more than 1,000 employess, 41 percent in companies of more than 5,000 employees (FASFID, 1977: 42). In such companies, production engineers do not design new equipment, and design engineers do not supervise workers, although they may oversee technicians and more junior engineers. Moreover, technical staff work has itself been specialized, even in the more traditional industries and companies. In the French context of the reality and ideal of *Monsieur l'ingénieur* such specialization and lack of line authority give rise to particular frustrations.

Yet, engineering appears to remain less specialized in France than in the United States, Japan, and Britain, probablay because of the lesser subdivision of lower and middle management in general (Granick, 1972: 298). Granick explains this difference and its significance as follows:

The disadvantages of specialized knowledge without an awareness of the needs of surrounding functions is counteracted in the American environment by three phenomena: by providing a good deal of information concerning the division and company generally, by the organization of considerable lateral cooperation with other managers in other functions, and by the transfer of managers among functions and administrative levels of the company. For French managers, none of these phenomena exist. Thus, in order for them to carry out their jobs properly, their positions must be defined more broadly than seems typical in American organizations. Nor are the costs of such broad definition exorbitant, since the losses from on-the-job training can be amortized over a relatively lengthy period of time.

Granick's comment refers to such aspects of French engineers' jobs as the problem of communications in French organizations and the tendency for French managers and engineers to stay in their jobs considerably longer than their American counterparts. On the other hand, they move up to lower managerial positions a good deal faster than Japanese graduate engineers. The latter often spend the first ten years of their careers doing a series of the kinds of jobs done by French technicians or even skilled workers, accumulating in the process a *polyvalence* that is much more based on knowledge of concrete details than is true among the more abstractly *polyvalent* French (personal communication from Hiroatsu Noharu). In Britain, the craft tradition appears to inhibit such *polyvalence*, and leave technical workers divided into many more specialized technical occupations (Smith, 1987: 80–92). This book returns to these matters in chapters 6 and 7. First though, it is important to examine the organization and nature of technical work at PAMPCO and TELECO.

Engineering at PAMPCO: organization and activity

There are many similarities in the organization of engineering work at PAMPCO and TELECO, but there are also significant differences. To understand them, it is necessary to examine each firm in some detail. At the

PAMPCO site, research, technical services, and production-maintenance coexist in three entirely separate and autonomous divisions. Each of these broad functional areas is further divided into a number of separate departments that specialize in some particular product, process or problem. Consider the Technical Services Division; its organization chart is displayed in Figure 5.1. The smaller figure at the right end of each department box represents the number of *ingénieurs* (*cadres*) in that department. The larger figure represents the total number of personnel in the department. Virtually all those who are not *ingénieurs* are *techniciens*, including some *techniciens supérieurs*.

It is evident from this chart that the *ingénieurs* are scattered about several rather specialized areas that have little to do with one another. Even within the design (*Etudes*) department, each of the four teams is responsible for developing different types of technical improvements for different aspects of the production process – the big pipe-spinning machines in the first case, other kinds of production machinery in the second, etc. Consequently, each team can operate quite independently of the others, and there is little need for coordination among them.

There is further specialization within these teams. For example, of the four *ingénieurs* under the *chef de service* of the *Equipements* team, one has specialized in the problems of heating and cooling since coming to PAMPCO eleven years earlier. A second, also at PAMPCO eleven years, has always worked in the same department (*Etudes*) and is "a specialist" (according to the Division Director) in metal-casting equipment. The other two also specialize in particular problems. Specialization here does not entail much interdependence with fellow team members. There must be considerable coordination with the equipment users in the factory, but that relationship resembles more closely one between professional advisor and client than one among members of a coordinated team of interdependent specialists. Since these PAMPCO specialists need not be responsive to changes in the work of their fellow specialists, and since PAMPCO's products and production process hardly change from year to year, it is possible to define quite carefully the different responsibilities involved in each engineer's job. This accounts to a large extent for the existence of the detailed job descriptions at PAMPCO.

It is also evident from the organization chart that there are about twice as many *techniciens* in most teams as there are *ingénieurs*, and that PAMPCO takes the distinction between *cadres* and *techniciens* (*etam*) seriously enough to list separate totals for each in the lower left hand corner. These *techniciens* are normally assigned to specific *ingénieurs*, and this has implications for the place of both in the division of technical labor. For the *ingénieur* it means that he is a manager of sorts as well as a technical staff specialist. More importantly, it allows him to assign routine tasks to his

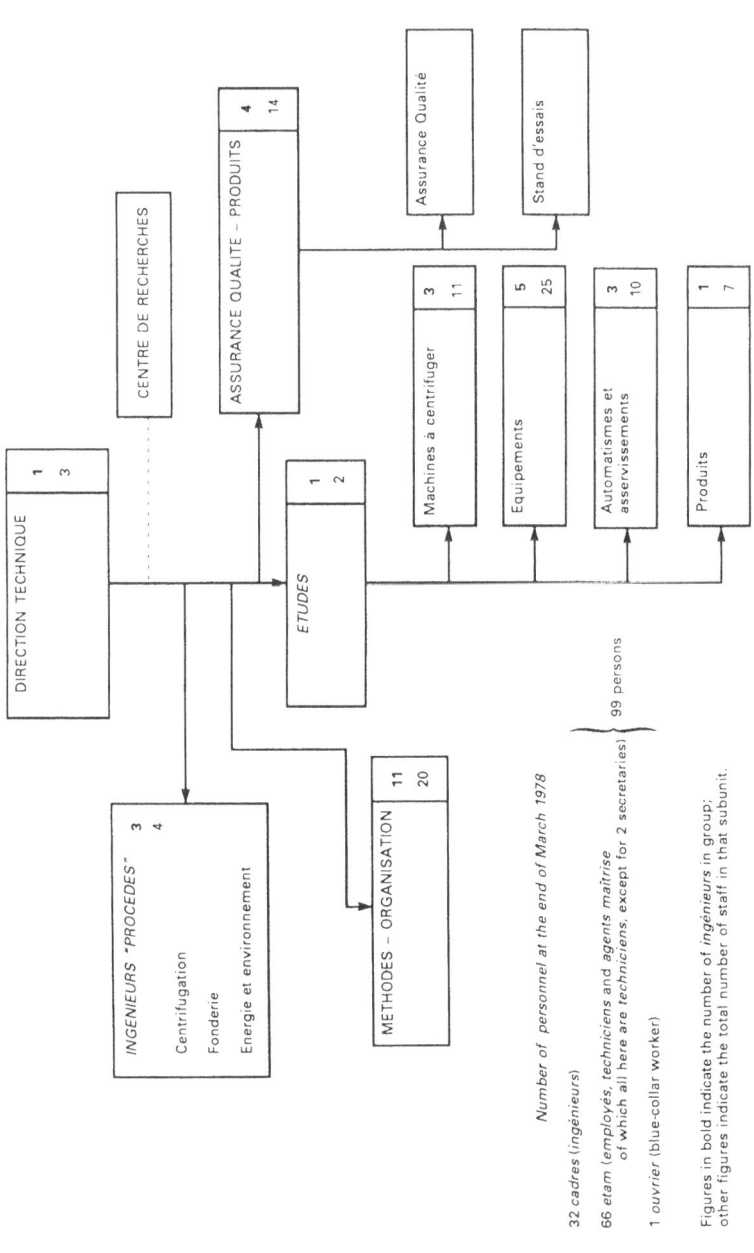

DIRECTION TECHNIQUE

| **1** |
| 3 |

CENTRE DE RECHERCHES

INGENIEURS "PROCEDES"

| **3** |
| 4 |

Centrifugation

Fonderie

Energie et environnement

METHODES – ORGANISATION

| **11** |
| 20 |

ASSURANCE QUALITE – PRODUITS

| **4** |
| 14 |

ETUDES

| **1** |
| 2 |

Machines à centrifuger

| **3** |
| 11 |

Equipements

| **5** |
| 25 |

Automatismes et asservissements

| **3** |
| 10 |

Produits

| **1** |
| 7 |

Assurance Qualité

Stand d'essais

Number of personnel at the end of March 1978

32 *cadres* (*ingénieurs*)

66 *etam* (*employés, techniciens and agents maîtrise*
of which all here are *techniciens*, except for 2 secretaries) ⎫ 99 persons

1 *ouvrier* (blue-collar worker)

Figures in bold indicate the number of *ingénieurs* in group;
other figures indicate the total number of staff in that subunit.

Figure 5.1: Organization chart, PAMPCO's Technical Services Division

technicien. PAMPCO's research division is organized quite similarly, and again there are about 140 *techniciens* assisting 60 *ingénieurs*. Production engineering involves more supervision of manual workers, and is less specialized in some respects, but is also divided into a variety of areas of responsibility.

In short no two *ingénieurs* at PAMPCO do the same job; judging from the organization chart, it would seem that the *métier noble d'un ingénieur* has undergone considerable rationalization. In order to interpret the significance of such specialization however, it is important to examine more closely the work these *ingénieurs* actually do.[3]

For this purpose, official job descriptions can be helpful. Detailed job descriptions exist for all engineering positions at PAMPCO except those in research. Illustrative is that of Etienne Soulet (pseudonym), a 36-year-old *ingénieur diplomé*, an *Ingénieur II* at PAMPCO, and the *chef* of the little *automatisme groupe*. This *groupe* includes three *techniciens supérieurs* and four ordinary *techniciens*, as well as two other *ingénieurs*.

According to Soulet's job description, he and his *groupe* are responsible for designing and carrying out various projects for further automating PAMPCO's local production machinery. More specifically, the Soulet team's assignment is to "establish the specifications, design the new devices (*réalise les études*), order the material, manage the bookkeeping (*gère les crédits*), oversee the construction of new devices, help with the installation and test runs of machinery equipped with new devices, train the users, and provide ongoing technical assistance to users." The job description does not specify particular activities, but it does indicate the nature of the coordinating he should do with suppliers, users, and other parties. For example: he should "maintain frequent contact with the suppliers of relevant electronic components in order to remain informed about available techniques."

Soulet finds this written description of his duties quite accurate, simply emphasizing that his group sees a project through from beginning to end, from initial design to testing of the finished product. When asked how he spends a typical day, he explains that it depends on the stage of the project. During the design stage, he is usually at his desk – one of several in the large design room. Later, he is in the factory overseeing the installation, testing, and refining of new automation devices. At all stages, his workday involves a fairly wide variety of specific tasks, the sequence, pacing, and exact nature of which are controlled by him and continually changing.

Thus, a job that appeared highly specialized at first glance turns out to involve a wide variety of technical, administrative, and managerial responsibilities. True, the two *ingénieurs* who worked under Soulet did more specialized jobs, but it is important here to distinguish between jobs and persons, and to keep in mind the career mobility of *ingénieurs*. Soulet himself once did one of these jobs. His two assistants were less likely to enjoy such promotion – indeed, they had been in the same positions for several years – but they were *ingénieurs maison*, former *techniciens* or *ouvriers professionnels* for whom the current job represented upward mobility from more routine jobs.

Now consider the work of a fairly typical maintenance engineer in the PAMPCO factory, Bernard Barrier. He is also an *Ingénieur* II and also *diplomé*. According to the job description, he has no other engineers under his command, but he does supervise seven technicians or foremen and 73 workers. Like most engineers who are not department heads themselves, he reports directly to the head of his department. Monsieur Barrier and his subordinates are responsible for the maintenance of six pipe-spinning machines and the accompanying equipment (for feeding molten iron into them and for finishing and removing the pipes). Because this machinery is critical to the production of the factory's main product and operates twenty-four hours a day, the maintenance team is under intense pressure to keep it running properly and to repair it rapidly whenever there are problems. Maintaining it requires an understanding of, in the words of the job description, "several different technologies: mechanics, hydraulics, electronics, etc., whose continual evolution further requires a capacity to keep learning and to adapt accordingly." The job description goes on to indicate this engineer's general responsibilities for maintaining the machinery, analyzing breakdowns, and developing less costly solutions to maintenance problems, advising design engineers working on modifications of the machinery, and supervising, training, and motivating his own subordinates. It instructs him to meet daily with his department head and his own assistants.

What this job description does not convey is the climate of pressure, tension, and interpersonal conflict typical of such maintenance work. The production engineers, themselves under pressure to meet output deadlines, tend to blame the maintenance staff for breakdowns and to press for rapid repairs. The maintenance engineers complain that it is impossible to repair things both correctly *and* rapidly, especially in view of their limited means, and they resent the way they are treated. The fact that both production, and maintenance engineers – especially the latter – work long hours and are frequently summoned to the factory at night and on holidays adds to the strain of this work. One does not have to spend much time in the factory to observe a heated argument.

Monsieur Barrier's typical day follows a somewhat more predictable pattern than Monsieur Soulet's. Beginning at 7.30 a.m. he first examines the report of the night shift and then investigates the problems reported. From 10.00 a.m. until 11.30 or 12.00, he attends the daily meeting with the production engineers. There he and the other maintenance engineers give their status reports, listen to the complaints and requests of the production engineers, and participate in the discussion and decision-making about solutions. After the meeting he discusses with the technicians and workers the work to be done, makes assignments, and orders any needed parts or tools. After lunch he checks periodically to see how the work is progressing, providing technical advice, general coordination, and administrative support as needed. Between these on-site checks, he works at his desk, examining technical documents, talking on the phone to PAMPCO design engineers or to technical representatives of the companies that construct the pipe-spinning machines, ordering supplies, calculating costs, and writing reports. He is interrupted frequently.

One of the features that distinguished Monsieur Barrier's work from that of Monsieur Soulet, is the pattern of "rounds" it involves, a pattern observed by Zussman (1985: 104–105) among American engineers. Barrier spent much of his time out of his office. Usually he was either attending regularly scheduled meetings with the production engineers or checking up and assisting with the repair work in progress, but sometimes he was responding to a maintenance emergency. Even when in his office, he was engaged in a variety of irregular activities, from calculating costs and writing reports to arguing with PAMPCO design *ingénieurs* and ordering supplies.

Similar "rounds" play an even larger part in the work of most production engineers. During the day I spent with production engineer Marcotte, he was on the move much of the time between the 8.30 a.m. and the 5.30 p.m. meetings of the *cadres* and foremen in his department. An older *ingénieur maison*, his job is to review the various foundry pieces recently rejected by Quality Control, diagnose the problems, and initiate corrective action. It took an hour just to inspect physically the several bins of rejects scattered about the large plant. Monsieur Marcotte personally examined the cracks, chips, and holes in various cast iron pieces, often marking them with coloured chalks. He then located the quality control foreman and politely argued with him about some of the rejects that Marcotte thought satisfactory. Unable to agree, they went together to the head of Quality Control. There, all three men discussed the technical issues involved, while passing selected pieces of the rough, heavy iron back and forth for examination. Once they had reached a compromise, Monsieur Marcotte turned his attention to locating the exact source of the production imperfections. Moving from one point to another in the production line, he inspected the speed, temperature, smoothness, and consistency, of the liquid metal, the sand casting molds, the machinery, etc. Twice he went back to his small, cluttered office to look at blueprints and technical specifications, stopping once to consult with the one design technician assigned to this production department. It was mid-afternoon before he felt confident enough to advise certain engineers and foremen as to the adjustments that needed to be made in the rather complex and delicate process of casting iron to production specifications. Most production engineers have more supervisory responsibilities than Monsieur Marcotte, but their days typically involve similar trouble-shooting rounds.

Thus, it is clear that despite the extensive division of technical labor at PAMPCO, the *ingénieurs* themselves did jobs that involved considerable variety, autonomy, and responsibility. It remains to be seen, however, whether the work of engineers in the more advanced industrial setting comes closer to that predicted by the proletarianization theorists.

Engineering at TELECO: organization and activity

The organization of engineering at TELECO resembles that at PAMPCO in several ways, but what is most interesting are the ways in which it differs. Consider the Simulator Division. Research and production personnel as well as those in design are under the same *Directeur Technique*. Only the sales engineers report to a separate *Directeur Commercial*. However, this integration of design and production is deceptive, for the manufacturing of simulators is essentially an exercise in design, and the few engineers who do work in production, testing, and maintenance work out of specialized departments within the Technical Division. The Technical Division is divided into two major groups, project coordination and design-production. Project coordinators are further specialized by type of project. The design–production group consists of eight different departments, including one for production, one for maintenance and testing, and one for drafting. Most of the engineers work in one of the three specialized development–design departments: electronic systems, mechanical equipment, and computer software. The organization of the much larger Telephone Division is more complex, but similar in most respects.

To understand how this organization of technical work differs from that at PAMPCO, it is instructive to compare one of the Simulator Division's design departments to the Technical Services Division already examined at PAMPCO. At first glance, the electronics systems department reveals a series of groups and teams specialized by product or problem, much as with the design department at PAMPCO. Closer inspection, however, reveals that these Simulator Division teams and the engineers in them all work on projects which are parts or aspects of a single but complex and interrelated system, a simulator. Thus, the design engineers must coordinate their activities with those of many other design specialists, both within and out-side their department. As a result, they have less control over their projects than their PAMPCO counterparts have over their smaller but more inde-pendent ones. On the other hand, their individual work takes on heightened significance for its obvious contribution to the functioning of a large, impressive, and constantly changing product.

The existence of special "project coordinators" at the Simulator Division

Table 5.1. *Percentage of time spent in various activities by design ingénieurs diplomés, by firm*

Activity	PAMPCO	TELECO
Meetings	7.5	6
Contacts outside department	13.0	6
Technical – conceptualization	25.0	46
Technical – execution	10.0	9
Technical writing	8.0	9
Administration	6.0	2
Relations – subordinates	12.5	6
Relations – colleagues	6.5	11
Relations – superiors	11.5	7
	100.0%	102.0%

facilitates this coordination of technical activities among different specialists. The result is somewhat less communication with engineers outside their own department than found among PAMPCO design engineers. This is evident in the differences between the average "time budgets" of the six PAMPCO and eight TELECO *ingénieurs diplomés* in design.[4] Table 5.1 reveals that the TELECO design engineers spend considerably more time than their PAMPCO counterparts actually designing solutions to technical problems, and correspondingly less on several other activities, including contacts with people outside their department.

This time-budget analysis also shows that TELECO's *ingénieurs d'études* spend considerably more time than their PAMPCO counterparts on one type of activity, designing solutions to technical problems, and correspondingly less time on several other types of activity. It would be a mistake, however, to interpret this concentration on "conceptual" activity as evidence of greater specialization, much less deskilling, at TELECO. It is important to distinguish between specialization regarding activity and specialization regarding area of knowledge. Some professors engage in research, writing, and consulting as well as teaching, while others only teach. Yet those who are strictly teachers often teach several different subjects, while those who do more than teach may be quite specialized in one area of expertise. In the case of TELECO specialization in conceptual activity is associated with variety and change in the type of problems on which *ingénieurs* work, partly because of the rapid changes in the technology and products.

These time-budgets also reveal significantly different patterns of relations with colleagues, subordinates, and superiors. The TELECO engineers have considerably less contact with superiors and subordinates and somewhat more contact with their colleagues in the department. These patterns reflect differences in the proportions of ranks at the two firms. In the Simulator Design Department there are ninety-six *ingénieurs* but only half as many *techniciens*. Moreover, most of the *techniciens* work in large testing laboratories apart from the offices of the *ingénieurs*. This means not only less communication between *ingénieurs* and *techniciens* than at PAMPCO, but less of a command relationship between them and less opportunity for *ingénieurs* to assign their more routine tasks to their *techniciens*. This was especially true in the Software Department, and generated some complaints among the engineers there, a point to which I return below.

The engineers here do both the systems analysis, which is real engineering work, and the programming. The latter, which takes a lot of our time, is the work of *techniciens*, but here it's done by *ingénieurs* because they already know the programs well having done the analysis. The engineers in the electronics department have more responsibility, more subordinates, and more contact with clients. They are more typical engineers.

Examination of the actual work done by TELECO *ingénieurs* throws additional light on these observations. To begin with, there were no job descriptions at TELECO, for the important reason that in such a changing industry work assignments evolve rapidly in response to developments in technology and customers' demands. As in many small-batch, high technology industries, production and maintenance are minor functions employing very few engineers. Most of the engineers do one type or another of design work, although there are also small contingents of project, production (including methods and maintenance), testing, sales, installation, and service engineers. In both the telephone and the simulator divisions, the design engineers are divided into several specialized areas, the two largest of which are "software" and "hardware." Software engineers design, write, test, and rewrite computer programs intended to meet the specified communications needs of particular simulator or telephone exchange projects. These engineers work in offices of several similar engineers, each one at a desk that he (or occasionally she) leaves only to work on a computer terminal (often with a technician) in a separate programming and program testing lab or to attend a meeting. They appear to work steadily but without urgency, taking the time now and again to chat with office-mates about sports or politics.

Hardware specialists – recall Pierre's description of his work in chapter 3 – design the electronic circuitry that communicates the instructions in the computer programs designed by the software engineers. They choose the appropriate electronic components (transistors, condensers, memory chips, etc.) and work out their appropriate configurations and connecting circuits. The design depends considerably on the particular capacities and pieces of the various components, many of which are new and untested items supplied by outside firms. As a result, these hardware engineers have more contact with suppliers, with the test technicians in nearby laboratories, and with the methods engineers. Indeed these contacts and the responsibilities they represent imply more administrative work – i.e. writing reports at one's desk – than was the case for the software engineers. In short, the hardware engineers are less specialized – at least in some senses of the term – than their colleagues in software, and have more authority and a broader spectrum of contacts with outsiders; that is, their work resembles that of many PAMPCO design engineers more closely than it does that of many other TELECO design engineers.

For all these TELECO design engineers, however, the daily work varies considerably, depending on the current project and their stage of progress on it. Asked to describe their typical day, most said that there was no such thing; their work activities varied too widely. At the most, there was a loose cycle of sorts that extended over a period of many months. First, they were engaged in studying a problem and planning their approach to it, then designing a specific subsystem and carrying out calculations, then testing and revising, and finally writing a report. All along the way there were meetings and consultations with others, including suppliers, clients, and fellow engineers working on other aspects of the same larger project.

There are other kinds of design engineers at TELECO, but their work and daily routines do not differ greatly from those of the hardware engineers. By contrast, the work of sales engineers is quite different. Not surprisingly, they spend a great deal more time communicating with clients and potential clients – in person, by telephone and by letter. They also engage in a considerable amount of internal communication with their own superiors and with project managers. Simulators and telephone exchanges are large, complex and expensive pieces of equipment that are highly customized to the particular needs of specific clients. (At PAMPCO, the *cadres commerciales* are not even engineers.) Selling such equipment means not only negotiating and writing technically and financially complex contracts, but also representing the client to the company and vice versa during the months or even years of development. Delays and modifications are common and often require explaining, costing, and contract renegotiating. The sales engineers do most of this work, and in the process find themselves doing a fair amount of technical-commercial writing – of proposals, contracts, letters, memos, and reports. They also organize and conduct visits, tours, and demonstrations for clients or potential clients, set up special training programs, and do a number of other similarly administrative tasks. Because so much of this work involves delicate relations with major customers, it is closely monitored by the *Chef de Service Commerciale*. A sales *Ingénieur* II with several years of experience in the department must still clear with his supervisor each letter he sends out. Thus, while the work of these engineers is neither specialized nor routine, it involves no authority over anyone except a secretary and quite limited autonomy.

Little need be said about the production and maintenance engineers at TELECO, for their work closely resembles that of their counterparts at PAMPCO. The "methods" and "industrialization" engineers, however, do somewhat different types of work, and there are more of them. This reflects the project character of production at TELECO. Consider Monsieur Jalbert, an *ingénieur maison* at the Telephone Division's factory in the Western provinces. He is in charge of a small team that is responsible for determining the cheapest method for constructing a proposed piece of equipment, and, if the project is approved, of then overseeing the implementation of the cost-efficient techniques. His team, which includes a systems analyst, works closely with two other teams, one consisting of experts in computers and computer languages, the other of engineers and technicians who specialize in the production and testing of hardware.

A typical day for Monsieur Jalbert begins with a glance at the incident report from the previous night. He immediately investigates the reported trouble spots or breakdowns, which means a visit to a development room, test lab, or shop floor. The duration of this activity varies widely, depending on the problem. Once he has taken whatever immediate action he can, Monsieur Jalbert returns to his office and analyzes the results of various practical experiments his team is conducting – a calculation of how long a way of doing something takes, for example. This is followed by a tour of the development lab to check on various aspects of the progress on the team's current project(s). In the afternoon, he may spend a good deal of time talking with the representatives of the potential user of the device, with the suppliers of parts, or with the members of the two other teams. He also answers inquiries about this project and the team's availability for undertaking new projects. Towards the end of the day, he answers his mail and works on written reports. In this work, he does not have to go to company headquarters in Paris to consult with design engineers very often, but some of the "industrialization engineers" at his factory go as often as once a week. In all of this, we again see the phenomenon of the engineering "round," but in the case of this innovative industry, rounds for some engineers are combined with an aspect of technical work associated only with design engineering at PAMPCO, unique projects that come, evolve, and give way to unique successors.

There remains a type of engineering work that deserves special mention: research. At PAMPCO, research is more applied, experimental, centralized and bureaucratized than at TELECO. At PAMPCO there is a *Centre de Recherches* where teams of engineers and technicians work at adapting PAMPCO's traditional product, pipes and valves, to new or expanding markets – the transportation of hot liquids or high pressure gases through "hostile" climates, for example. These teams often work on a contract basis for another division in the company, trying to improve some specified performance characteristic within certain cost limits. Much of the work involves setting up and monitoring experiments, and writing up the results in a research report. Most of these research staff are fairly recent graduates of prestigious *grandes écoles* or are highly qualified technicians; there are almost no *ingénieurs maison*. In most cases an *ingénieur diplomé* has his own small project and one to three technicians to assist him. He works out of his own office; the technicians (many of whom are also engineers, by our definition) stay in the laboratory. He has considerable autonomy with respect to his daily schedule, but he does face project deadlines and has little choice about the projects on which he works.

Research at TELECO is much less centralized. Within each division (product line) there are a few engineers whose primary job is to think about new products or processes. At the Simulator Division, there is one engineer who works full time at trying to think of and design new applications of electronic simulation. At the Telephone Division, the *Service Technique* includes, along with its dozens of software and hardware design engineers, a team of three that could be called one of research engineers. Working alone at their desks without any assistance from technicians or others, these engineers are attempting to think up new applications of advanced computer technology to telephone services. At the time of the research, one engineer was concentrating on extending the "numerical logic" of the "temporal" (as opposed to the "spatial") telephone switching system to consumer equipment as well as central exchanges. According to him, effectiveness in this work required a good understanding of the market and trends in it. "If the French Telephone System (P.T.T.) shows an interest in further development of our idea, then the team will take on a technician or two and build a feasibility model" (pre-prototype). And if the project is further approved for development and testing of a prototype, it is quite likely that these research engineers will move with it, working on and perhaps even overseeing each stage from design and production to installation and testing.

Research, then, is less specialized at the electronics firm, and more integrated into general production activities, as Burns and Stalker (1961) have argued is appropriate for innovative industries. Indeed at the Simulator Division, applied research constitutes a portion of the work of several "design" engineers. Each new simulation project (often resulting in only a few simulators) is unique, and often involves new developments in the technology of reproducing the desired sights, sounds, and motions.

In summary, by the standards of the traditional *Monsieur l'ingénieur*, the design engineers in both firms do somewhat specialized technical jobs. Yet, it is important to be more specific. First, the degree of specialization is hard to quantify, for conceptual as well as methodological reasons. There is no disputing that today's engineer works on a smaller fragment of the technical problems involved in producing a final product, but these problems have grown in number and complexity. Second, specialization itself is an ambiguous and multidimensional concept. A technical job may be limited to a small aspect of the production process – automating machinery or developing a new visualization technology – but that assignment may involve a wide variety of technical, administrative, and other activities, and may require using knowledge of several engineering specialties.

Third, to judge by PAMPCO and TELECO, the nature of specialization varies between traditional and high technology engineering firms in a way that may be consequential for experience, relationships, and forms of solidarity. This difference is similar to Durkheim's (1947) famous distinction between segmented and interdependent divisions of labor.[5] In an interdependent division of labor, there is specialization within the group, and solidarity evolves not out of likenesses among self-sufficient individuals, but rather out of complementary differences. At PAMPCO, there was specialization, and, in the final analysis interdependence, but specific engineering jobs tended to have a discrete, segmented character about them that permitted the engineer to work on his tasks in a relatively independent manner. By contrast, at TELECO, the engineers worked on parts of larger projects that required more on-going coordination with other engineers. The time-budget analysis presented in Table 5.1 shows the TELECO design *ingénieurs diplomés* spending considerably more of their time – 11 vs. 6.5 percent – in relations with other engineers of similar rank, despite the existence of project coordinators. Such interdependence raises problems of control and coordination, and complicates performance evaluation by results. This may be one reason for assigning more of the technical work involved to *ingénieurs* rather than *techniciens* than is necessary on grounds of skill. In general, *ingénieurs* and *cadres* are more trustworthy and can be relied on to exercise initiative and responsibility in ways that serve the company's goals.

Fourth, the qualitative examination of the work several engineers do reveals considerable heterogeneity within the occupation of engineering and a fairly wide variety of tasks usually associated with any one engineer's job, this in spite of our ignoring *techniciens supérieurs* and *chefs de service*. Only the software design engineers at TELECO appear to work at quite specialized jobs.

Fifth, the qualitative investigation also shows the project nature of much engineering work, a characteristic which inhibits routinization and has implications for engineers' sense of time, purpose, and achievement. It is noteworthy that the one group for which work was not project-oriented were the production engineers at PAMPCO. Their work was not very specialized, but there was much routine in their daily schedules of meetings, "rounds," and office work, which hardly varied day after day and month after month. By contrast, the projects on which the TELECO design engineers work appear to be the most technically challenging and rapidly evolving.

Sixth, most of these engineers have considerable latitude in deciding exactly how and when to perform their assigned tasks. With the qualified exception of PAMPCO's production site and TELECO's sales department, immediate supervision was absent. Nor, with the possible exception of the software engineers, did the organization of the work and work flow control task activity. Pressures to perform well are built into the reward system, but there appears to be considerable autonomy on the job.

Finally, there is the question of the level of skill itself. This is extremely hard for a non-engineer to judge, and I shall rely largely on the engineers' own judgments, expressed in the next section. First though, I want to address Bauer and Cohen's argument about the codification of engineering know-how. As suggested in chapter 4, there may be some codification of the theoretical knowledge that engineers apply, but the result is simply that engineers remain engineers rather than becoming applied scientists. Bauer and Cohen (1982) make much of the manuals to which some *ingénieurs* and *cadres* refer, but their examples are of non-technical *cadres*. Moreover, it is important to distinguish between manuals that routinize work and those that do the opposite. It is not only repairmen that use manuals, so do scientists and physicians. These manuals often reduce the amount of tedious work to be done by providing tables of information that previously required laborious calculations. Social scientists need only think about computing the chi square statistic by hand to be reminded of the enormous conveniences afforded by some standardization and codification. In any case, it is my impression that few of the manuals used by engineers instruct them on what to do and/or how to do it. Company procedures do guide the flow of activity, as does the prior organization of work, but the *ingénieurs* I interviewed valued these as devices that helped them find the specialists they needed to consult when they needed them. In the words of one *ingénieur diplomé* at TELECO in answering the question about the sources of the knowledge he used: "It is my experience that counts, especially my knowledge of the operating procedures in a big company; that's important for coordination; I consult with specialists a lot, so I must know where to find the appropriate ones."

Other authors have made similar points with regard to the mechanization of technical work. For example, Smith (1987: 187) observes that the British aerospace design engineers he studied "did not, on the whole, consider their conditions of labour would be eroded by CAD (computer assisted drawing)." Even the draftsmen, who make extensive use of mathematical calculations and geometry in their work, considered the pocket calculator a useful tool. In Smith's words (1987: 146), "No one expressed alarm at the calculator removing or reducing arithmetical skill. A drawing machine in the office, used mainly for tape control drawing, was not considered a threat to skill either."

This analysis of the actual work done by several engineers suggests that although they do quite specialized jobs at any one point in their careers, they enjoy, especially in France, much more task variety and job autonomy than recognized by most proletarianization theorists. Nevertheless, it remains to be seen how the *ingénieurs* themselves view such specialization, and what their attitudes reveal with respect to the deskilling debate.

The meaning of specialization

Several of the interview questions attempted to elicit the engineers' own feelings about their work. Asked whether they got to use their technical skills as much as they would like, similar majorities of engineers in both firms replied in the affirmative. However, among *ingénieurs diplomés* alone (with sales engineers excluded), 37 percent of those at TELECO replied in the negative, as compared to only 22 percent at PAMPCO. By contrast, among the *techniciens supérieurs*, more at PAMPCO than at TELECO felt underemployed. These are not large differences, and it would be a mistake to draw weighty conclusions about deskilling, but they are compatible with the earlier observation that PAMPCO's *ingénieurs* are in a better position to delegate simple tasks to *techniciens*.

Each engineer interviewed was also asked whether his own job was varied enough, or, on the contrary, entailed too much routine work. Only 8 percent at PAMPCO and 7 percent at TELECO described their jobs as routine or boring; 63 percent at PAMPCO and 59 percent at TELECO said they were varied enough (or even too varied), while the remainder gave such intermediate answers as "sometimes it's routine, sometimes not," "it's beginning to get dull," etc. However, the location of these slightly bored engineers was significantly different in the two forms. At TELECO it was the engineers who worked in the big pools of design engineers who were most likely to complain of *la routine*, while at PAMPCO it was the relatively unspecialized engineers in production.

In considering how specialization may affect the work experiences of engineers in science-based industries, it is important to bear in mind some of the disadvantages of traditional production engineering jobs, unspecialized though they may be. A PAMPCO *ingénieur maison* put it well:

My job is relatively varied. We touch a lot of areas: electronics, computing, hydraulics, mechanics, electricity, organization, management, the environment, etc. If one wants, there are even a good many possibilities for little reseearch projects. But nevertheless, the work remains routine in the form of the daily program I gave you. It's like that practically every day.

[Interviewer: Do you think it's better in the Design Department?]

Yes, their problems change, and they have more contact with the outside. For us the outside arrives via the Design Department, and they have already made many choices before bringing things to us. They make most of the decisions. That's why the Design Department is more appealing.

Another production engineer – an *ingénieur diplomé* explained the problem of *la routine* in this way: "At PAMPCO the production process is well understood – it's very routine . . . and that I don't like." He went on to say that specialization within limits is a very good thing, and gave the example of Monsieur Soulet's Design Section, which specializes in electronically automating certain production machinery. He felt that such specializing gave an engineer an opportunity to pursue a problem in depth, and that the more specialized technical staff jobs often provided two things woefully absent in production, opportunities to be creative and from time to time new projects.

Even such boredom as was expressed by this production engineer must be interpreted carefully, however, for the normal careers and career concerns of engineers influence their experience of what might otherwise be negative characteristics of their jobs. In the words of the production engineer quoted above:

I still don't regret transferring to production, because that has allowed me to gain promotion more quickly. The work is more constraining and dirtier, but it has enabled me to advance my career faster than if I had stayed in design.

Chapter 7 examines engineering careers in detail. The point here is simply that in engineering, variety can be repetitive while specialization may mean quite interesting and challenging work. Part of the reason for this is that specialized jobs tend to be in research and design, and research and design work tends to be project-oriented by nature. Projects come, evolve, and are completed, to be replaced by new and perhaps quite different ones.

Even within evolving projects, the degree of specialization varies, and movement to more specialized work can be gratifying (rather than a price paid for involvement in innovative, high technology projects). Consider the following reply to the question: "Can you briefly describe your current job, including its responsibilities and what a typical day is like?"

I work on the definition, development, adjustment (*mise au point*) and documentation of the [technical term] subassemblies of the MT 20 [an electronic telephone exchange]. I do less *mise au point* now that the team has grown and some specialization has emerged. But I don't miss the broader range of activities; testing was really the work of a *technicien* – though good experience. My job involves some real conceptual work, though less now than at the beginning, when we only knew the performance objectives, not how to accomplish them. [*ingénieur diplomé*, TELECO]

This project nature of certain engineering jobs is characteristic not only of particular functions, but also of entire industries, especially high technology industries. In contrast to PAMPCO's stable, mass-production technology, the production at TELECO's Simulator Division is in very small batches highly customized to the client's specifications. Furthermore, the technology is evolving rapidly. Several engineers expressed awareness of these aspects of their industry in replying to the question: "are some engineering positions too specialized to be interesting?"

No, I don't think so. It is very rare, but it may happen. Here it doesn't because of our type of product. It might happen in assembly line production. [a *chef de service*]

No, perhaps because we don't mass produce our product (*fabrique en grande série*) ... And you can change from one function to another here if you find work becoming less interesting. [an *ingénieur maison*]

No. The technique is evolving so fast that there is always something new to learn. [an *ingénieur maison* in production]

If project-type work and rapidly evolving technologies keep even *some* specialized engineering interesting, the opportunity to pursue a problem in depth is characteristic of *only* specialized positions.

No. There is some tendency towards specialization, but that means more opportunity to go into depth. [*ingénieur diplomé*]

No. I was once the most specialized at TELECO in my department, but it wasn't all that specialized, and I found it very interesting. Specialization opens new doors for those who are able, new opportunities to go deeper into one's own specialty or related areas. [*ingénieur diplomé*]

There are other desirable aspects of specialization. As one TELECO engineer explained, if one's technical tasks are interesting, it is better to spend a great deal of time on them than to be distracted by paperwork, meetings, and problems with subordinates. "For me, repetition is not a problem because I like my work." Several engineers mentioned how specialization implies interdependence and thus the satisfactions of teamwork and of seeing how one's personal job contributes to the larger whole. Production and maintenance engineers also interact with "others," but these "others" tend to be subordinates and superiors in situations made tense by time pressures and authority differences. And as Crozier (1964) has observed, the French seem to find face-to-face authority relations particularly stressful.

Other evidence also suggests that job content was not a significant source of dissatisfaction at either company. Each *ingénieur* and *technicien* was asked: "Do you plan to remain at PAMPCO/TELECO?" and "why?" Although more TELECO respondents (25 percent) than PAMPCO respondents (12 percent) expected to leave their firm eventually, the distributions of reasons for leaving were virtually the same at both firms, with pay and career reasons predominating, and only 18 percent mentioning elements of job content.[6] Moreover, among the vast majorities expecting to remain with their firm, 80 percent at TELECO mentioned liking their work as a reason, as opposed to 47 percent at PAMPCO.

Each engineer was also asked if there were any morale problems among the *ingénieurs* in his firm. Whether because they like their work or discount their dissatisfactions in the expectation of a promotion out of it soon, these engineers expressed little concern about work content; only 9 percent at each firm mentioned it. Asked which of their job activities gave them the most satisfaction, 86 percent at each firm said their technical activities, as opposed to administrative, managerial, or other activities. This is not to say that they complained only about the organization hassles and paperwork. At TELECO in particular the *ingénieurs diplomés* complained about some of their more routine technical activities, such as testing or lower level programming. Yet most of them seemed to view these the way professors view grading exams and papers, as an unfortunate aspect of an otherwise good job.

Techniciens and *petite école ingénieurs*

Although many engineers regard some quite specialized jobs as fascinating and some quite unspecialized ones as boring, the fact remains that at TELECO some engineers in specialized jobs regarded their work as either somewhat or very routine. Moreover, about half thought that some engineering jobs had become so specialized as to be uninteresting. In six cases they were referring only to the jobs of *techniciens* (five of these were *techniciens supérieurs* asked only about "*techniciens* like you"), but the remaining eighteen referred to *ingénieurs*. Many pointed to those in software who designed new computer programs, saying such things as "specialization is making such work repetitive and too limited; can't see the whole" and "once you have mastered the specialty there is nothing more to learn and the job becomes dull." Some of the engineers who replied "no" to the question (about some jobs being so specialized as to be dull) quickly added that the specialized jobs would become uninteresting after a while if one could not rotate to a new position. Indeed it seems that one major difference between those answering "yes" and "no" was the degree of confidence about the ease with which an engineer could transfer. Some said anyone who had initiative or kept up with the technology could change; some even claimed it was company policy to rotate engineers. Others were more skeptical:

Programming is much too specialized (*parcialisé*). It's interesting for about one and a half years, but what the firm's interested in is repeating the system because it's more profitable, and so the job becomes repetitive, uninteresting. [*ingénieur diplomé*]

There's little movement between departments because it's hard to change from working on software to hardware. [*ingénieur diplomé*]

As with so many things, those engineers with superior credentials found it easier to obtain transfers. *Techniciens* and *ingénieurs maison* faced not only discouraging prospects for promotions, they were also in a weak position to bargain for lateral transfers. As one *technicien supérieur* explained,

Work becomes repetitive, even in a design office, once you know your area. That takes several years in *électro-mécanique*, less in software. The systems and projects change but basically they are just variations of familiar problems. It's the same for *ingénieurs* but they can change departments or advance more easily and faster.

The relationship between *techniciens* and *ingénieurs* is an important one and, as previously indicated, it differs significantly from one company to the other. At PAMPCO very few *ingénieurs* complained about specialization, but many *techniciens* did. The reason is that once a job was so specialized and simplified as to be boring, it was assigned to a *technicien*. Indeed, that is the basic logic of deskilling any job: simplify it enough, often by means of dividing a complex whole into simpler parts, to hire someone less skilled and therefore cheaper; Braverman (1974: 79–81) refers to this as Babbage's principle. But in the case of the engineers it is a logic that may protect as well as threaten the occupation. Many engineers may escape deskilling and proletarianization through the assigning of the more simple of their tasks to the *techniciens* who assist them. Whole functions may be relegated to the *techniciens*. According to one maintenance engineer at TELECO, "Eventually maintenance will not be interesting or require *ingénieurs* because repair will be just a matter of changing cards and modules." Yes, even most *techniciens* emphasized that their dissatisfaction is with blocked mobility, not the work they do. This is consistent with Fossati and Said's (1982) findings from a survey of ninety-three *techniciens supérieurs* (53 BTS, 40 DUT) that 80 percent liked their work but 56 percent felt frustrated about the prospects for promotion and their exclusion from decision-making.[7]

However, at TELECO some thought that *ingénieurs* were doing work that was routine and unchallenging. At the Simulator Division, two or three of the youngest *ingénieurs diplomés* complained that "we are only technicians here." However, this word "technician" is an ambiguous one, and must be interpreted with care. Recall Pierre's remark (p. 42): "Now the technical sophistication of the equipment and products requires a greater number of engineers who have become, in effect, technicians. None of their other assets and qualifications, such as management and personnel supervision, are called upon as they were years ago." Here, it is evident that by "technician," Pierre does not mean lower level technical work, but rather highly specialized technical work that involves no managerial activity.

Nevertheless, a few TELECO engineers did seem to think that the company was intentionally hiring engineers to do the work of *techniciens*, and that was why there were many discontented young *ingénieurs* at TELECO. But why would a company hire more expensive workers to do work that could be done by less expensive ones? Parts of the answer emerge from the replies of two experienced TELECO *ingénieurs* to the question about some jobs being too specialized.

Interviewer: Do you think there are some engineering positions here which are so specialized as to be uninteresting?

For sure, particularly in software. We use *ingénieurs* almost like semi-skilled workers. That creates many problems. In the software team there is extremely high turnover.

Interviewer: But nevertheless you hire engineers [for these jobs] because of a need for a high level of technical competence?

Well, I have the impression that it's much more seamy (*beaucoup plus sordide*) than that. We hire *ingénieurs* because even a naive young *ingénieur* is nevertheless very bright; he can be put to productive work more rapidly. That short-circuits the training period. If one were to use inferior workers, there would be a much longer training period . . . In my opinion, it's a misguided policy; it damages team morale, and that is serious.

This engineer was a *chef de service* who thought technological developments would eventually solve the problem by freeing software engineers from the more tedious aspects of calculating programs and enabling them to "recover the noble work of analysis and creation." A second *ingénieur* had a somewhat different view:

Yes, work can become very repetitive, depending on how the work load falls. Such a bad division of work is forced by the rush to meet orders. The company hires engineers to do the work of *techniciens* because *ingénieurs* are cheaper now [had mentioned surplus of *ingénieurs* earlier]. The *ingénieurs* from today's *petites écoles* are comparable to the *techniciens* of an earlier era. It's not a question of technology, but of the educational system and company policy and interests.

These two answers shift the explanatory focus from the logics of technology and capitalism to the structures of particular product and labor markets. Within France TELECO operates much like a defense contractor. Because sophisticated telephone exchanges and electronic simulators are manufactured by extremely few firms in the world, because the major purchasers are governments more interested in performance and speed of delivery than price, and because the French government tends to "buy French," TELECO is often under greater pressure to produce rapidly than to produce cheaply. The changing nature of the market and the product put a premium on a flexible, *polyvalent* work force as well. In these conditions, TELECO evidently finds it advantageous to use *ingénieurs* in jobs that could be done by *techniciens*, but probably not as effectively over time.

The more general and important point here is that firms may compete in a variety of ways besides price; they may compete through product development, product design, product quality, delivery dates, after sales service, etc., and employers have a certain amount of autonomy in choosing their strategies, contrary to the market determinism of neoclassical economics. Labor economist Jill Rubery (1987: 7) makes the point well:

Empirical and theoretical work in the labour process debate have revealed . . . [that] [f]irms are concerned with all aspects of their competitive position and may be willing to concede even wage increases as rewards for the implementation of new technology if this would ease the transition and enable them to improve their position in the product market. Moreover, as product variety and product development have been found to be taking on increased importance in mature industries as well as new industries, the implicit assumption in Braverman that firms are competing primarily on least cost methods of producing standard commodities does not hold.

Moreover, employers such as TELECO may be buying something more than a higher level of skill when they hire *ingénieurs* instead of *techniciens*. They may be hiring those who can be more completely trusted to work responsibly without supervision. Responsibility may be particularly important in high technology industries characterized by extensive reciprocal interdependence. Interdependence among specialists creates two problems for management. The one most often noted is coordination. It greatly simplifies matters for management if it can get specialists to coordinate themselves. This means allowing them a good deal of discretion about when and whom to consult. Permitting discretion is not risky if the consequences of abuses of it or of mistakes are unimportant – i.e. inexpensive, as in the case of a floor sweeper, but, as Whalley (1986) correctly emphasizes, in complex, high technology projects, discretion becomes more problematic.

This leads directly to the second problem for management, the evaluation of results. Evaluation by results may prove an adequate substitute for close supervision in some circumstances, and surely gives management some control over the role performances of engineers. However, in situations of high interdependence, it can be impossible to evaluate any one individual's contribution to the success or failure of the team's product. It is for work that is both difficult to supervise or evaluate individually, and for which the consequences of errors are expensive, that management seeks – must seek – trustworthy workers. From management's point of view, *ingénieurs* (but not *techniciens*) are fellow officers who have indicated their reliability by making it through an *école* and investing in a job whose key feature is the long-term career rewards offered in exchange for the faithful discharge of responsibility.

Finally, the existing labor market situation in France in the late 1970s favored a tactic of substituting *ingénieurs* for *techniciens*. The acute shortage of high level *techniciens* and the emergence of some unemployment among *ingénieurs diplomés*, made the latter relatively cheap in the short run (see Table 7.5, p. 151). This situation was rooted in particularly French institutions and practices. Historically France has suffered a serious shortage of highly trained industrial technicians. As late as 1967, Gilpin (1968: 351) could write: "If France has a serious scientific management manpower problem, she has a national crisis in the shortage of engineers and technicians . . . Actually, the situation is more acute with respect to 'technicians'than 'engineers'. Germany, for example, produces about three times as many technicians per year as does France."

It was for this reason that the French government established the Instituts Universitaires de Technologie in the late 1960s. But the IUT's failed to live up to the hopes for them (Boudon, 1977). The reasons for this continuing shortage are too complex and debatable to discuss here, but they involved elements of social structure and industrial development specific to France, including the abhorrence of specialization.

While the shortage of *techniciens supérieurs* persisted (despite impressive growth, especially in the broader category of *techniciens*), the production of *petite école ingénieurs diplomés* increased dramatically (see Table 4.1, column C). More specialized and easier to obtain admission to, these *petites écoles* are in many respects producing the functional equivalents of *techniciens supérieurs*, but by law and custom the *diplôme d'ingénieur* puts their graduates in a different category than *techniciens supérieurs*. And by long-standing traditions in French industry, their credentials entitle them to begin their active careers as *cadres*. The literature on labor markets increasingly recognizes that custom and the patterns of expectations it generates are influential forces for the maintenance of internal labor markets long after their economic rationale has disappeared (Doeringer and Piore, 1971: 40–41; Osterman, 1984: 6, 179; Althauser, 1987). Thus, just as the shorter forms of higher education are, in Boudon's words, "a bad bargain," the *petites écoles d'ingénieur* appear to be a good bargain.

Yet, good bargains seem less of a bargain if too many people seek to take advantage of them. Chapter 7 shows that the career payoffs for *petite école ingénieurs* are less than for their *grande école* counterparts. Moreover, 59 percent of the interviewed *ingénieurs diplomés* at TELECO were graduates of *petites écoles*, as compared to only 35 percent at PAMPCO (although the percentages of elite *grandes écoles* were 26 percent and 20 percent respectively). It was *petite école ingénieurs* that TELECO seemed to be recruiting to do the "work of *techniciens*," though it is unclear whether such recruitment reflected choice or lack of choice. It may well be that *petits ingénieurs* are or will become less trusted, that the association between *ingénieur* and trusted worker will weaken, and that the structuration of the boundary between trusted and untrusted workers in France is changing. If a shortage of *techniciens* generates pressure for internal promotions at the same time that the gap in the schooling between certain strata of engineers and technicians is converging – due to the emergence of both *petite école ingénieurs* and *techniciens supérieurs diplomés*, this would seem to weaken the *cadre* barrier. This in turn would help account for the relatively greater job satisfaction of the *techniciens supérieurs* at TELECO, by comparison to PAMPCO, the relatively lower satisfaction of the *ingénieurs diplomés*, and the continuing presence of *ingénieurs autodidactes* in high technology industries.

Even if some *petite école ingénieurs* are doing work that requires less skill *and* responsibility than that required in normal engineering jobs, it is important to be cautious about generalizing to the whole occupation of engineering. It probably is true that some high level technical work is always in the process of being routinized. If that work is eventually assigned to *techniciens* and the proportion of *ingénieurs* in the work force continues to grow, the net effect is still one of skill upgrading. But even if a new stratum of lower *ingénieurs* emerges to do some of that work, thus reducing the average skill level of all *ingénieurs* considered together, it is still premature to conclude that deskilling has occurred. It is critical to keep in mind how the occupation has grown, how there may be just as many *ingénieurs* as in the past doing highly skilled engineering, and that they may have been supplemented rather than displaced by a new lower stratum. In France, this stratum consists of people who probably became *techniciens* in the past, but today become *petits ingénieurs*. One disadvantage of such inflation of educational credentials is that high-level training no longer reaps the intrinsic and extrinsic rewards that it once did, and individuals experience the disappointment of frustrated expectations for higher payoffs.

Attitude measures at one point in time are not conclusive evidence against a deskilling argument, but the data presented above do suggest that most of the work done by most engineers at two major French firms has not been reduced to the "formatted routines" imagined by proletarianization theorists. Moreover, a recent Lou Harris-France poll of 400 1977–78 graduates of engineering schools specializing in electronics, *électrotechnique*, and *informatique* finds 97 percent of them fairly or very satisfied with their occupational experience since graduating (*Nouvelle Usine*, 1982). One explanation for the resistance of much technical work to deskilling is that it is too unpredictable to routinize. It is unpredictable because ongoing changes in technologies and markets mean frequent changes in either products or the processes for producing them. This is especially true of the high technology industries, which is a major reason they employ so many engineers. Furthermore, when high level technical work is routinized, it is assigned to *techniciens*, or increasingly, to *petits ingénieurs* who represent in many respects contemporary functional equivalents of them.

Conclusions

The first section of this chapter pointed to the traditional role of a French engineer as a technical manager and to the importance of *polyvalence* for French engineers' exceptionally successful claims to both managerial careers and high social status. The next two sections described the responsibilities and daily work of selected *ingénieurs* at PAMPCO and TELECO, emphasizing the variety of administrative and technical tasks involved in any one job and the contingent character of much work. In the case of production and maintenance engineers, work is organized into "rounds" which entail a broad range of technical and managerial activities, but which do not change much from day to day. In the case of most other engineers, work is organized around evolving "projects," and the variety of tasks is more evident over time than across space.

These two sections also called attention to the ambiguities of the concept of specialization, and distinguished two different forms of it. Characteristic of stable, old industries, or at least of those producing large batches of small, technically simple products, is a segmented division of technical labor that frees engineers of much routine work – it is delegated to *techniciens* – and leaves them quite independent in carrying out a small project. The other form of specialization, an interdependent division of labor more typical of high technology industries like TELECO, does not necessarily entail greater specialization, but does require more coordination among individuals working on various aspects of the same large, complex, and evolving project. Coordination and subsequent adjustment require a high degree of initiative, responsibility, and adaptability, qualities which may be enhanced by training in general science, but which largely reflect character rather than knowledge. These are qualities of the "trusted workers" produced by French engineering schools.

In the fourth section, the analysis shifted to the attitudes of the engineers themselves. Were they unhappy with the content of their work, and if so was the reason one that supports the proletarianization argument? In general, most *ingénieurs* expressed satisfaction with the work they do. Only 8 percent of the interviewees at PAMPCO and 7 percent at TELECO viewed their jobs as routine or boring. Complaints about under-utilization of acquired skills were more common, especially among PAMPCO *techniciens* and TELECO *ingénieurs diplomés*, but still not widespread. In answer to a question about morale problems, only 9 percent at each firm mentioned the content of their work tasks.

Specialization seemed to have little bearing on the distributions of these sentiments. Some specialized jobs seemed to be valued for providing opportunities to pursue a problem in depth, to be creative, to change work with the evolution of a project, to avoid the headaches of supervising and paperwork, and to enjoy the satisfaction of teamwork. Other specialized jobs, especially some in programming, did seem to be insufficiently challenging and varied. The analysis of this phenomenon emphasized the advantages at TELECO, given its market situation and the supplies of technical labor, of assigning some *ingénieurs* to some jobs which involve a certain amount of "technicians' work." It also acknowledged that the argument that routinized engineering tasks get assigned to technicians means that there are normally some engineering jobs at the margin, jobs still being done by engineers but soon to be delegated to subordinates.

Finally, this chapter in general, and the fifth section in particular, have offered some explanations for the finding that deskilling is not a serious problem for French *ingénieurs*. This concluding section extends that discussion. First, engineering by definition involves the use of technical knowledge to solve industrial problems, and problems tend to be by nature non-routine, even for experts in certain types of problems. Second, changing technologies and markets continue to stimulate a steady flow of challenging technical problems, and it is hard to imagine an end to that flow in the foreseeable future. Third, and related to the first two points, much engineering work is project-oriented; and it is hard to routinize such work, for projects vary and evolve, especially in the new industries. Fourth, although there may be cost-cutting reasons to routinize some high-level technical work, firms compete in other ways than price, and need not continually try to rationalize all labor. Moreover, as trusted workers with long-term careers at stake, *ingénieurs* simply do not withhold labor – neither to minimize effort nor to protect their jobs. Indeed, they may have a built-in incentive to work hard, for their jobs and promotions in the future sometimes depend on the success of the project being worked on in the present, and part of the definition of success is meeting a customer's deadline for an estimate or prototype.

Fifth, while some engineering jobs are quite specialized in at least some senses of that ambiguous term, few engineers are. The reason is that engineers' careers propel them through a succession of jobs that are often quite different from one another and carry the engineer out of purely technical work and into management. The *polyvalence* of French *ingénieurs* facilitates such mobility. Moreover, the engineers at PAMPCO and TELECO had considerable say about their job assignments, so many who are specialists are so by choice.

It may be true that some deskilling is taking place at any given moment within the world of engineering work, but two other things are occurring at the same time. The deskilled work is being assigned to lower level technical workers, and new types of technical problems (or expanding demand for existing types of problem solvers) are creating new engineering jobs and even specialties at or near the top of the skill hierarchy. Even within the high technology industries, the numbers of engineers are growing faster than numbers of technicians; thus, Groux (1985) notes that between 1976 and 1982, the numbers of *ingénieurs* and *cadres* in the electric and electronic industries (including telecommunications) increased by 2.9 percent, and the numbers of technicians by 2.4 percent, while the numbers of workers declined by 6.2 percent. In view of engineering's place in the skill hierarchy and the large growth in its ranks, it is hard to imagine that the undocumented declines in average job content alleged by proletarianization theorists are not more than offset by the "compositional shifts" (Stark, 1980).[8]

Highly skilled engineering jobs probably account for as high a proportion of the work force now as thirty or fifty years ago, but they have been supplemented, not replaced, by a new lower stratum of slightly less skilled – or at least less managerial – engineering jobs. At the same time French engineers, even those graduating from prestigious *grandes écoles*, are facing growing competition from business school graduates for the top management positions in French industry, as chapter 7 shows. These developments – an expanded bottom tier and more limited career possibilities for those in the top tier – may well explain some of the frequently expressed anxiety about the decline of the French engineer. Yet such developments in the position of French engineers reflect less the logic of capitalism than the social definition and organization of French technical workers and their careers. In the meantime, as many practicing French engineers as ever continue to experience the intrinsic job satisfaction of high-skill engineering.

6

Autonomy and authority on the job

As employees in bureaucratic organizations, engineers are subject to supervision, evaluation, rewards and punishments by those in positions of authority over them. Central to the recent theoretical debate about technical workers are claims about the character and legitimacy of that authority. Theories of proletarianization predict a tightening of supervision and control. Theories of professionalization and a new working class share an expectation of growing demands by engineers for greater autonomy and/or participation at work, especially in the science-based industries. In addition, the Marxists, the neo-Marxists, and Daniel Bell all foresee an anti-capitalist undercurrent to the critique of industrial authority, while Freidson anticipates an emphasis upon occupational rights.

This chapter examines evidence bearing on these expectations in the case of France, a country that offers a particularly interesting site for the investigation of these arguments. On the one hand, the long history of anti-capitalist movements in France and the recent wave of interest in self-management should favor the development of the anticipated challenges to bureaucratic authority. On the other hand, the tradition of *polyvalent cadres* would seem to raise obstacles to either the proletarianization or the professionalization of *ingénieurs*. The first two sections of this chapter focus on engineers' experiences of autonomy and authority at work. Succeeding sections examine their experience of bureaucratic authority and attitudes towards it, demands for more participation in management, and needs and desires for authority. The final section considers attitudes towards capitalist authority – authority justified by property rights and market considerations.

The experience and bases of autonomy

Proletarianization theory and theories of professionalization that anticipate a crisis of legitimacy tend to assume that bureaucratic authority is fairly intense and visible. But is this true, or do most engineers enjoy considerable autonomy in the day-to-day performance of their jobs? And do they attribute any constraints they do experience to the exercise of bureaucratic authority *per se*, or rather to the imperatives of industrial production and marketing? Finally, if engineers do bump up against transparently bureaucratic authority, do they regard it as legitimate, and are those in the science sector more likely to question it than their traditional counterparts? In addressing such questions, it is useful to treat autonomy and authority separately. It is appropriate to begin with the former, for the experience of some job autonomy may affect reactions to commands by bureaucratic superiors. Thus, this section focuses in turn on the experience and structure of autonomy, while the following section examines the operation and perception of authority.

Freedom from direct supervision

With the exception of some *techniciens*, most French engineers experience very little direct supervision. By direct supervision, I mean control over how a job is done, through detailed instructions on how and when to do it and/or periodic inspections by the supervisor to make sure the job is being done properly. By indirect supervision I mean control by means of job assignments, requirements for approval of decisions (expenditures, etc.), and evaluations of performance results.

Judging from both observations and the reports of the engineers interviewed, direct supervision is rare. Engineers are normally assigned responsibilities, not tasks; and they are given considerable discretion in meeting them. This is evident in the written job descriptions examined in chapter 5, as well as in the detailed account of a day spent with Monsieur Marcotte, the production *ingénieur autodidacte* at PAMPCO. Monsieur Marcotte enjoyed great autonomy in conducting his inspection of the quality control rejects and initiating corrective action. Moreover, the constraints he faced came not from his own department *chef*, but from the engineers and the *chef* in another department – quality control – with which his department had a lateral relationship.

Some of the interview data on this issue suffer imprecision from the vague and overly-structured wording of one relevant question: "In general, does the control of your work and your results seem rather heavy, rather light or what?" Nevertheless, the replies are instructive. Only 6 percent of the respondents said that such control was too strict or rather heavy. Ten percent said it was sometimes strict, sometimes loose. The remainder said that control was either about right (18 percent), pretty or too relaxed (34 percent), exercised by results only (12 percent), or some combination of these. Although these replies reveal less about the frequency and character of supervision than about perceptions, it is reasonable to infer from such favorable evaluations that these engineers experience little direct supervision of their work.

Moreover, supporting evidence emerges in the responses to a less structured question about autonomy: "Do you think it would be desirable, in order for you to do your job better, to extend your decision-making freedom, your responsibilities, or the scope of your activities?" The wording and breadth of this question invite comments about indirect as well as direct supervision, so many more engineers answered in the positive. Nevertheless, 46 percent identified no managerial constraints, and the remainder focused essentially, whether critically or not, on indirect controls. The following are illustrative of the comments by those who identified no constraints.

I have plenty of freedom of decision, maybe too much. I choose my own materials, including the supplier of them, and order them myself. I could make all the mistakes imaginable because my superior isn't interested in these technical problems. [a PAMPCO *ingénieur*]

I have enough freedom of initiative and decision making – enough autonomy. I'm told the objective and the deadline and then I'm on my own. [a TELECO *ingénieur*]

It is not my boss who tells me: allot so much time to this task, so much time to do this, do that. Myself, I reckon I have a task to accomplish, and for this task I organize my time according to the results I want to obtain. Of course, I must attend the meetings which are organized because it's important that we all meet regularly. But aside from those, I organize my schedule as I please. [a PAMPCO *ingénieur*]

In reply to the question about the desirability of greater decision-making freedom, some engineers did identify managerial constraints, and a good many were critical of them. However, these constraints involved indirect controls, especially the obligation to obtain permission from a superior before initiating a change or expenditure, and are examined below. More relevant to the present point is the *absence* of references to *direct* supervision in answers to the question: "Would it be desirable to extend your decision-making freedom, etc.?"

Yes, a little, even though I have a lot of responsibility and power. There are some decisions I could make by myself which I have to refer to higher-ups. [a PAMPCO *ingénieur*]

I have a lot of responsibility, but not too much. The problem is that our superiors don't know what our problems are. It would be better if they gave me even more responsibility. [a TELECO *ingénieur*]

Yes, more freedom to invest [in new equipment]. I appreciate the amount of autonomy I have, but I feel cut off from higher level decisions. And I'd like more authority to promote [my subordinates] and give raises. [a TELECO *chef*]

Only a handful of *ingénieurs* mentioned excessive managerial inter-
ference in the carrying out of their daily duties, and even these comments
were mild.[1]

In certain areas the higher *chefs* are a little too involved in what we do. They do not
understand our results and sometimes ask us to do things that are unnecessary. [a
PAMPCO *ingénieur*]

Yes, if I were free to do so I would do my job differently. My *chef* tends to be
traditional in his idea of authority but he is beginning to loosen up. [a TELECO
ingénieur]

In view of the observational data and of these responses to the two questions
about autonomy, it is evident that most of these engineers enjoy considerable freedom from direct supervision.

The structure of autonomy

Indeed, there are at least three good reasons for viewing such autonomy on
the job as structured into the organization of technical work.[2] The first is that
engineers often develop specialized expertise that their supervisors lack.
This makes it difficult for their supervisors to tell them how to do a particular
job. On the contrary, supervisors often find it in their interest to ask subordinates whether something is technically feasible, what additional
resources would be needed, how long it would take, and whether there
might be a "better solution" to the problem. In such a situation, supervisors
often endorse decisions reached by consultation and consensus, rather than
impose their own. What the engineer experiences is not the authority of
office but collegial relationships in which influence reflects relative levels of
expertise.

To the extent that French engineers are more *polyvalent* than American
engineers, at least by training, individual engineers are less likely to be
expert in an area about which others know little. In this sense *polyvalence*
simultaneously weakens the position of the subordinate and strengthens that
of his supervisor. On the other hand, while many French engineers continue
to obtain *polyvalent* technical educations, most work at somewhat
specialized jobs. The longer they remain on one such job, the more expertise
and thus autonomy they are likely to acquire. While specialization may
make a job more boring, it may also contribute to the autonomy of the
specialist.

The responses to the question about control support this claim. None of
the interviewees who described the control of their work as strict were
specialists in the sense of spending 40 percent or more of work-time on one
activity. By the same token, *techniciens*, specialists who stay in one job a
long time, were much less likely than *cadres* to perceive the controls as strict,
while *chefs de service*, managerial generalists, were most likely to report

controls as strict. The finding appears at odds with Bailyn's (1980) suggestion that the job satisfaction of engineers rises with increases in administrative responsibility because higher administrative jobs entail greater autonomy.

The second reason for viewing a certain kind of autonomy as inherent in the organization of technical work is that most engineering jobs require considerable freedom of physical movement. Consequently, it is hard to supervise engineers directly, whether experts or not. Every *ingénieur* at PAMPCO and TELECO had an office or a desk in an office, but spent a good deal of time elsewhere. Only 28 percent of the interviewees spent more than 80 percent of their work day in their offices; almost half spent 50 percent or less. Of the few who spent virtually all their time in one place, whether office or not, most were *techniciens*. Some of the time spent outside offices was spent in consultations and meetings with other engineers. Much of it was spent in laboratories or on factory floors. Like Monsieur Marcotte, production and maintenance engineers moved about while checking on supplies, machines, workers and output, and attending to the numerous small problems that plague production lines. Research and design engineers inspected experiments and tests being run by laboratory technicians, while research, design, and production engineers visited each others' offices, shop floors, and labs to collaborate on the need for or operation of some new piece of equipment. Supervisors rarely accompanied the engineers during these meetings or inspections, nor did they try to authorize or monitor them. It is far easier and cheaper to allow engineers to organize their own time, and then judge them by the results of their work.

A third source of freedom from direct supervision is the predominantly lateral flow of communications among technical workers. The refined division of technical labor means that at any one moment, most engineers are performing but one of the many operations involved in the complex and expensive transformation of ideas and raw materials into finished products. Such work requires a considerable amount of coordination and mutual adjustment across specialties. Thus, unlike specialists in academia or medicine, engineers operate in a world of interdependence and interaction across departments and functions. Even at PAMPCO, with its more segmented division of technical labor, many requests for work – for different materials, temperatures, speeds, tests, applications, producedures, etc. – flowed laterally across and between departments rather than vertically through the chain of command. As long as such requests were reasonable the engineers were expected to honor them without making their initiators go through the supervisors. Work assignments thus flowed not down from one supervisor, but rather in from many colleagues.

At TELECO, the pressures for lateral communications were even greater, for there a variety of specialists in programming, electronics, and mechanical systems were at work on the same complex and internally coordinated machine. That machine was often a prototype for which the specifications evolved with progress in the solving of new problems in specialized areas. It is true that the lateral communicating was somewhat institutionalized in the role of project coordinators, especially in the Simulator Division. Nevertheless, outside the software departments most TELECO engineers participated in more lateral coordination than did their counterparts at PAMPCO.

As Zussman (1985: 107–108) notes, such a lateral flow of work assignments affects the engineer's experience of authority in two important ways. First, to the extent that requests for work come from many colleagues, not all of whom can be quickly accommodated and none of whom has direct authority over the engineer, the engineer gains a certain amount of leverage for bargaining about what work he will do, and when and under what conditions he will do it. At TELECO the existence of more than one "boss" is actually formalized in that firm's "matrix" organizational structure. Formally, each engineer reports to two or more *chefs*, one of whom is the head of his specialized functional unit, while others are the project coordinators of the projects on which he is currently working. In fact, most engineers spend little time with any of these *chefs*; rather they respond to the requests of many of their co-workers on the same project.

Second, the lateral flow of communications and work requests obscures the role of bureaucratic authority and even of human agency in the management of work. Laterally communicated work requests tend to be visibly linked to, and thus justified by, the requirements of developing and producing a certain product. Here, however, it is necessary to distinguish between the two firms. At TELECO, and in many such "R&D" companies, each project is tailored to the specifications of a particular client. Throughout the two or three year life of a project, the client's technical representatives participate in the developmental work, and often request product changes in response to unforeseen developments in technology, costs, or the client's needs. Work requests to a particular engineer stemming from a client's request for product changes can thus be legitimized by the demands of the market. Such legitimacy, of course, does not imply autonomy, but it gives constraints a natural and necessary character. With respect to the discretionary authority of management, the experience remains one of autonomy. At traditional PAMPCO, by contrast, a paternalistic style of management, especially pronounced in the production–maintenance division, meant that higher managers could and occasionally did interfere in the work of junior *cadres*, sometimes by means of forceful commands.

It follows from this analysis of lateral communications that a sense of freedom from direct supervision should be even more pronounced at TELECO than at PAMPCO. It is the case that all four of the engineers who complained that control was too (or "rather") strict worked at PAMPCO. Moreover, 71 percent of the TELECO engineers characterized the controls as quite or too relaxed, while only 43 percent of PAMPCO engineers agreed. Whatever the explanations for these company differences, it seems reasonable to reject claims of growing tensions between engineers and managers over direct control of technical work. Indeed, some TELECO engineers felt that controls were inadequate; they wanted less responsibility, or more evidence that their work was important enough to supervise, even if that meant more direct supervision of their own work.

Thus, the engineers at both firms, but especially those at the science-based one, experienced considerable freedom from direct supervision. Zussman (1985) and Whalley (1986) report similar findings about the American and British engineers they studied. More recently, Meiksins and Watson (1987: 9) write that of the 585 engineers they surveyed in Rochester, New York, 91 percent say that they are "usually or always able to decide how to go about their jobs, while 70.6 percent are usually or always able to obtain the resources they need without getting permission." Such evidence of autonomy is impressive, but hardly surprising if *ingénieurs* are viewed as trusted workers.

The operation and perception of bureaucratic authority

Although direct supervision was minimal, indirect supervision was both more widespread and more resented. By indirect supervision is meant the three types of intermittent controls mentioned above: supervision by assignments, approvals, and results. In these forms, bureaucratic authority did set limits upon the autonomy of the engineers, was visible, and was *sometimes* the source of serious grievances. This section examines the various ways that bureaucratic authority shaped the work lives of engineers at PAMPCO and TELECO, and analyzes these engineers' reactions.

Indirect supervision

Supervision by assignments refers to the actual organization of work – the creation of positions, definitions of jobs, allocations of individuals to them, and distribution of particular projects to particular engineers. At both PAMPCO and TELECO, *chefs* at various levels of the hierarchy made such decisions; they assigned engineers to positions and new jobs to engineers. On such occasions bureaucratic authority became a visible source of constraint. Nevertheless, it did so in ways that were acceptable to most of the engineers.

Asked how much choice they had about the projects to which they were assigned, most of the *ingénieurs* (i.e. excluding *techniciens supérieurs*) responded that they had some influence over such decisions. A few thought the choice was entirely their own; many said "within limits" or something similar. Some said they could express a preference, but were skeptical as to how much good it would do. The majority, however, appeared to share the following opinion of a methods *ingénieur diplomé* at PAMPCO:

People don't know about the up-coming studies until they are distributed, but you can let your *chef* know your taste on types of projects. He will try to accommodate you because it is in his interest.

Such an attitude suggests much less constraint and imposition than implied by proletarianization theorists. Only among the *techniciens*, especially those at PAMPCO, were there several negative replies.

No, never have a choice of projects. I just carry out orders, and that's all. [*PAMPCO*]

Never. It's imposed. Sometimes we can discuss a little, but it rarely matters. In general the *cadre*, the *ingénieur* with us, he knows our area and he distributes the work among us. [*PAMPCO*]

Not all the *techniciens* perceived so much constraint, and among those who did, there were some who expressed much less resentment. The following comment reveals a common rationalization at all levels for a lack of choice: "fairness."

No choice, not at our level. But that's OK. It's fair because if the project is not attractive to me it wouldn't be to others either. I'm more concerned about being free to do the work that is assigned me in my own way. [*a PAMPCO* technicien supérieur]

Not only did most engineers have some influence over their assignments, or at least feel they did; they also accepted the limitations on personal preference. They recognized as reasonable the assignment of work on the basis of such efficiency criteria as expertise and availability. As one PAMPCO *ingénieur* put it, "Projects are given to those who know the area, or who seem ready, able, and suited." The very fact of specialization within engineering makes certain assignments appear logical. Here, however, it is important to bear in mind the *polyvalence* of French engineers and the weak connection between education and employment. At PAMPCO in particular, many *ingénieurs* arrived without much prior specialization, and as young *cadres* in a paternalistic old company they had little choice about their initial assignment. The specialized knowledge that they soon developed was not in areas of their own choice, and while their *polyvalence* in training facilitated arguments for a desired transfer, it also made them more vulnerable to unwanted transfers and assignments.

Supervision by results requires little discussion. Almost all employed persons face the prospect of negative sanctions – firing, ultimately – for a job badly done, and most salaried workers in large hierarchical organizations face the possibilities of slower or faster promotions. The rewarding or punishing of engineers is facilitated by the visibility and measurability of the results of their work. Unlike the case with teachers or physicians, the fruits of an engineer's labors are embodied in manufacturing processes and physical products that are usually subject to direct evaluation in terms of quality and efficiency (Larson, 1977: 26–27). Interdependence complicates evaluations of individuals, but, in general, a supervisor need not stand over an engineer to know how he is performing.

However, visibility and measurability of task performance do not alone guarantee trustworthiness, or motivate employees to do their best. There must also be incentives to perform well. For salaried employees, the classic incentive is the possibility of promotions within the firm, assuming of course some supervisory discretion about granting them. Here again the authority situation of French engineers is both unusual and changing. Unlike in Britain, the higher management in France has traditionally been recruited from among *ingénieurs diplomés*, thus giving the latter better promotion prospects than their British counterparts (Sorge, 1979; Glover, 1983).

The possibilities of promotion to higher management, however, often allow and justify quite oppressive treatment of those who aspire to realize such career potential. This is especially the case in traditionally paternalistic settings. Paternalistic here refers to a relationship between organizational superiors and subordinates in which the norm is that of father–son responsibility and obedience rather than bureaucratically specified authority and obligations. Gallie (1978: 182) is correct when, referring to the refineries he studied, he writes: "broadly speaking, French management's strategy [is] paternalistic, while British management's [is] semi-constitutional." Yet, such paternalism is even stronger in old industry, and warrants further explanation.

Stearns (1978) attributes the rise of industrial paternalism in France to employers' needs to attract and retain workers in situations of labor shortages created by widespread peasant land owning and slow population growth. Labor shortages were especially severe in industries such as coal and iron that were necessarily located in remote areas. Reid (1985: 585) agrees with Stearns's analysis, but adds that "paternalism [in France] can best be understood as an effort to turn the worker into an employee." Reid points out that nineteenth-century employers faced a problem with labor contractors who exploited their workers and gave them no training. He adds, however, that "the foreman came to present many of the same liabilities as the contractor. In the last decades of the nineteenth century companies sought to concentrate not only technical power, but discretionary power as well, in the hands of the engineer, who by social origins and/or training was, unlike the foreman, clearly separate from the labor force." Of course, the model of the engineer as commander was already well established in the elite state engineering corps. Moreover, in a country that lacked a system of vocational education, on-the-job training was important, and engineers were better prepared than traditional foremen to instruct unskilled workers. Thus French paternalism can be understood as a labor market strategy. And as with many internal labor market institutions, aspects of paternalistic management endure long after the original rationale for them has disappeared.[3]

The rise of militant left-wing unionism, and the antipathy to collective bargaining among both unionists and employers in France reinforced paternalism (including company welfare schemes) as techniques of social control until the crises of the 1930s. With the massive strikes of 1936 and the Popular Front government's imposition of a national level union–employer agreement (the Matignon Accord), many employers gave up on paternalistic strategies for coopting workers. They did not give up on them, however, with regard to the emerging group of *cadres*. On the contrary, they applied them all the more strongly, especially in industries torn by labor strife. *Cadres* would be well treated, but in return much would be demanded of them, including complete loyalty and willing compliance to authority whose limits are often left to the discretion of superiors.

This distinctively French tradition is the key to understanding why the experience of supervision by results is becoming less, nor more onerous in the science-based sector. That such is the case is suggested by the responses to a question about what would happen if a superior was not very satisfied with the engineer's work. Only 24 percent of TELECO engineers indicated career or salary sanctions, as opposed to 57 percent at PAMPCO. The explanation seems to be the decline of overt paternalism among French managers. At old, family-dominated PAMPCO, promotions and bonuses were far more dependent on the favor of the superiors than they were at TELECO. As one PAMPCO engineer replied, "personality and personal relations play too large a role here." At TELECO, where there were no such complaints, promotions were more automatic; careers unfolded more predictably according to credentials and seniority. Consequently, there was less fear of negative sanctions for a poor job performance.

To be sure, there is also a technological reason for the milder supervision by results in the electronics firm. Stable, mass production techniques facilitate production control systems conducive to supervision by results. At PAMPCO, production engineers were confronted each morning with computerized reports on the previous day's output, costs, and inadequacies, and given new targets. Some of these engineers regretted the passing of production control from themselves or their department *chefs* to the technical staff at higher headquarters. One production *ingénieur* complained bitterly about the new production control system's restrictive impact on his initiative and responsibility: "We decide very little – we're almost reduced to just carrying out orders." The *chefs de service* in particular regretted their loss of authority. Some also doubted the accuracy and utility of the new system. But no one questioned the right of management to use it. After all, supervision by results is not particular to bureaucratic authority; it is implicit in any labor contract. A client who is not satisfied with the results of a contractor's work may threaten to take his business elsewhere if improvements are not forthcoming.

In short, supervision by results in French engineering seems to be less, not more oppressive in the science-based firms, both because such new firms are science-based, and because they are less likely to be characterized by the paternalistic practices of traditional French management. Moreover, supervision by results does not render bureaucratic authority problematic, and does help explain the absence of more direct supervision of technical work. There is one form of indirect supervision, however, that does make bureaucratic authority quite visible and problematic for many engineers, supervision by approval. This refers to the limits placed on an engineer's exercise of discretion in solving problems without the prior approval of his *chef*.

The world of engineering is one of ongoing change, of the development, testing and implementation of new techniques and products, often in response to changes in materials, technologies, and markets. Changes usually imply options – choices to be made among several possible solutions to problems. Ideas about solutions are cheap, and many employees offer them, but the implementation of an idea usually involves the commitment of organizational resources to it at the cost of alternative uses. It may also entail adjustments by others in the production system. Thus, it is hardly surprising that there are budgetary and other limits on the freedom of engineers to make such commitments; there are some such limits on virtually all organizational members. But in view of engineers' expertise in and responsibility for devising and managing technical change, it is also not surprising that some of them complain about the obligation to get approval for changes or expenditures they want to make in "their" areas of responsibility. Most of the positive answers to the general question, "would it be desirable to extend your decision-making freedom, etc.?" emphasized this constraint.

Yes: decision-making power and responsibility. I should have more authority to spend the money in my budget and more freedom to make repairs without getting permission from *chefs*. The bureaucracy is discouraging. [*PAMPCO*]

Yes, technical decision-making is too centralized and hierarchical. It's right that my *chef* intervenes in the definition of a project, but I should have more freedom of maneuver in the implementation. [*TELECO*]

Yes, in all areas . . . Not enough responsibility or decision-making freedom . . . We have to clear everything with him [The Division Head.] [*TELECO*]

These answers reveal the limits on job autonomy as experienced by some engineers, and comments like these were made by 39 percent of the interviewees. Another 14 percent identified similar limits, but were more accepting of them. In short, while engineers enjoy considerable freedom from direct supervision, many of them occasionally do bump against the realities of their subordinate status in a bureaucratic authority structure, and many engineers regard such supervision as excessive.

Again, it is at PAMPCO that a larger percentage of engineers report feeling over-supervised in this indirect sense: 45 percent as opposed to 34 percent at TELECO.[4] The main immediate explanation for this difference is the large difference between the *techniciens supérieurs* of the two firms. At TELECO only 33 percent of this category complained of excessive supervision, but at PAMPCO 68 percent did. The following quotes are illustrative:

Yes. Often we have the impression that solutions are imposed on us by the *cadres*. We have proposed better or simpler or faster solutions, but we are not listened to . . . We carry out orders, that's all. [*PAMPCO*]

No, we have a good deal of freedom to express ourselves as it is, and to make decisions. [*TELECO*]

Like the earlier findings on direct supervision, these comments support a view of PAMPCO as more rigidly stratified and stratum-conscious than TELECO. In particular, the distinction between *cadres* and *non-cadres* is taken seriously. As *non-cadres*, *techniciens*, including *techniciens supérieurs* are practically as well as officially outside the ranks of management. Only *cadres*, whatever their rank, wear green safety helmets: *techniciens* wear other colors, depending on their department. There is a directory of *cadres*, revised annually, that gives the age, educational credentials, current position, and address of every *cadre*, but only *cadres*. There is also a special lunch room for *cadres*. Such a *cadre salle à manger* also exists at TELECO headquarters, but only at PAMPCO is there also an attractive section of company housing set aside for *cadres*.

This paternalistic promotion of "*cadre* consciousness" is found frequently among old industry firms that employ a high proportion of manual workers. These are the firms that fear the defection of middle-ranking staff to the side of labor. Kolboom (1982) shows how employers encouraged the formation of the CGC, a relatively conservative and independent union of *cadres*, after the general strike of 1936 in order to preempt moves by *cadres* to join unions allied to the major workers' confederations. At the time of the research, PAMPCO still encouraged its *cadres* to join the CGC.

The limited autonomy of French *techniciens* stands in some contrast to Smith's (1987: 127) finding of a "high degree of job control and independence . . . in conditions relatively unfettered by either supervisory control or restrictive individual time constraints" among British technicians as well as engineers. Here again is evidence of the significance of distinctively national ways of organizing technical work and careers. In Smith's (1987: 156) words, "the lowly esteemed clerical, 'serving' or 'helping status . . . was not something that haunted the consciousness of British draftsmen, planners, and programmers.

At TELECO the traditional French boundary between *ingénieurs* and *techniciens* exists, but it is less pronounced. One likely reason is that there are too many *ingénieurs* at TELECO to treat them as members of a company elite. Moreover, the fact that there are twice as many lower level *ingénieurs diplomés* per *chef de service* in the electronics industry as in the metal-working industries means that the hierarchical distance between these two levels is greater in the science-based industry. In any case, higher management at TELECO does not encourage a "*cadre* consciousness," and as was evident in a few of the earlier quotes, some of the TELECO *ingénieurs* do feel excluded from management. In short, it seems that while the boundary between *techniciens* and *ingénieurs* has softened at TELECO, to the satisfaction of the *techniciens*, the distance between lower level *ingénieurs* and *chefs de service* has lengthened.

However, while this shift in the structure of authority appears to increase the autonomy of the *techniciens*, it is not clear that it does so at the expense of the autonomy of lower level *ingénieurs*. Only 28 percent of the TELECO *ingénieurs diplomés* complained of excessive supervision, as did 33 percent of their PAMPCO counterparts. True, some lower level *ingénieurs* at TELECO complained of being treated like *techniciens*. Moreover, fortified as they were by their numbers, their shared sense of isolation from *chefs*, a strong external labor market, and the absence of heavy-handed paternalism, they occasionally took collective action against management. For example, many young *ingénieurs* and *techniciens* demonstrated angrily at higher headquarters against their reassignment without consultation from their work place in the western suburbs of Paris to one in the northern suburbs. Yet, neither their concern about a decline in their status as *cadres* nor a readiness to display that concern means that these engineers suffer more indirect supervision than their higher status counterparts at a firm like PAMPCO. On the contrary, PAMPCO's paternalism and stable technology facilitated closer supervision by results.

In short, the engineers in the high technology firm enjoy greater freedom from indirect as well as direct supervision, largely because the management of their firm is less paternalistic than that of PAMPCO. PAMPCO engineers both complained about excessive supervision and feared negative sanctions more than did engineers at TELECO. No TELECO engineer made so strong a complaint about authoritarian management as the following remark by a PAMPCO design *ingénieur*:

I regret the amount of hierarchically defended violence by superiors in imposing their decisions . . . There's a military style of commanding practiced by some *chefs* here, and an inability to discuss differences calmly.

Yet, 56 percent of the PAMPCO engineers also reported that their company did *more* than the minimum for them as employees, while. none of the TELECO engineers agreed.

To the extent that French management is less paternalistic in advanced industries, engineers there have more freedom from indirect supervision, and enjoy it more. To the extent that French management in all industries remains personal, secretive, and authoritarian, relative to other countries, French engineers are likely to complain more about excessive supervision, and to take action against it where possible. In either event, the effects of ongoing industrialization interact with those of national institutions, in this case continuities and changes in French management practices.

In one sense, Mallet was right. The *ingénieurs* in the science-based firm are more militant than those in the traditional firm, and for reasons other than their proletarianization. Yet, working in France and within an intellectual tradition that tends to treat all capitalist societies as essentially alike, Mallet downplayed the ways in which French industrial institutions, especially paternalism and the *polyvalent cadre* tradition, lend a distinctively French tone to the impact of science-based industry. Moreover, he made assumptions about the radical character of *cadre* critiques of industrial authority that bear further examination. The following section examines in some detail the attitudes of the PAMPCO and TELECO engineers towards greater *cadre* and worker participation in management.

Participation and workers' self-management

The principal purpose of the discussion of indirect supervision was to reveal the limits of the engineer's autonomy and the objective structure of his authority situation. However, the criticisms of supervision by approval also indicated the existence of some discontent with bureaucratic authority. More specifically, the complaints about insufficient decision-making freedom imply demands for more participation in management, exactly the kinds of demands that Mallet anticipated in advanced industries. And demands for more participation may indeed, as Mallet argued, represent serious reservations about the legitimacy of bureaucratic authority. To get at these issues more precisely, the interview included two specific questions about participation in management, one about being more consulted in decision-making and one about industrial democracy. The following section examines in turn the responses to these questions.

The wish to be more consulted

Each engineer was asked: "Are there some decisions about which you think that you and the other *ingénieurs* [or *techniciens*] should be more consulted or informed?" If they answered in the affirmative – and 92 percent did – they were also asked what kinds of decisions, and why. At both PAMPCO and TELECO, almost 40 percent of the respondents (36 percent at PAMPCO, 39 percent at TELECO) expressed a desire to be better informed, but either did not mention consultation or explicitly rejected the idea of more of it. The following comments illustrate the most common reasons for rejecting consultation.

Yes, we need more information . . . As for participation by the personnel in decision-making, no, I distrust them.[5] They don't know all the parameters: financial, technical, etc. Not everyone is competent to make decisions. Better informed, yes, but not consulted. [a PAMPCO *ingénieur*]

I'm against schemes for more consultation. It will just waste time, make work, and soothe egos. [a PAMPCO *ingénieur*]

Better informed yes, but not more consultation. The more people who get involved in a problem, the more complex it gets and the more slowly one advances. Better to leave it to the competent few. [a PAMPCO *chef*]

We lack the knowledge, information, and thus competence to participate, and it would take too much time to keep informed. But they should tell us the reasons for certain decisions, and they don't. Even there though, one can go too far; too much information provokes endless disagreements. [a TELECO *ingénieur*]

Hard to say. I'm concerned about wasting time on things not of immediate interest to me. I have no real complaint here. I don't have time to get involved in and learn about everything, so must leave some matters to the experts. [a TELECO *ingénieur*]

These comments reveal a concern for efficient problem-solving, an assumption that most managerial problems are technical in nature, a belief that technical problems are best solved by those who are "competent" to deal with them, and a view of competence in any one area as concentrated among a few experts. Typically, these engineers did want to be consulted about matters in their own narrowly defined areas of technical competence, but also felt that they were already. Conversely, they felt that they lacked the competence to participate constructively in other matters, and assumed that those making such decisions were more qualified to do so. As one engineer explained it, such competence could be assumed because those managers had probably been assigned their responsibilities on the basis of competence, and because the particular job and experience in it positioned them to understand complications that would not even be evident to others. A secondary theme is that even if competence were widespread, decision-making by large groups is too cumbersome, divisive, and inefficient to be practical. In short, participation is not an end in itself but a means toward achieving organizational goals, and as such it should be limited to the few most competent. Not surprisingly, *chefs de service* were more likely than the other engineers to express such sentiments.

But what about the majority of engineers who did express a desire for more participation? Were not the values implicit in their complaints more radical? Although critical of the status quo, their demands for more consultation were limited, and the grounds of their arguments were strikingly similar to those of more satisfied engineers.

Yes, I am already consulted enough about most personnel and social policies, the main exception being promotions of my subordinates. But I would like to be more consulted about investments in new plant and equipment that involve my Department. I resent the fact that the decision-makers don't ask the advice of those who use the equipment. On the other hand, I'm skeptical about much *participation*, even by *cadres*, because we French talk forever. We exhaust ourselves in meetings. [a PAMPCO *chef*]

Yes, for sure; both [more informed and consulted]! Not for reasons of democracy at work, but because decisions would be better. I should be consulted more about the projects we undertake, the budget I have, and especially personnel problems. I would like to be able to recruit my own workers, interview them, and have more influence on their salaries and promotions. [a research *ingénieur* at PAMPCO]

On consultation, yes and no. No as far as such decisions as the selling of TELECO to XYZ, but yes for decisions which are often forced on us by other departments. There's more of a problem of coordination than autocracy. [a TELECO ingénieur]

Yes; decisions, for example, concerning the choice of a system . . . Often we are not *au courant* as to why one choice is made rather than some other . . . We could work better in the way desired by higher management if better informed . . . Work would be more pleasant, and we'd waste less time. [a TELECO *ingénieur*]

These engineers want to be consulted about matters which affect them directly, especially those on which they feel particularly competent by virtue of their specialized knowledge or organizational position. Most commonly, these are technical matters in which they are or will become directly involved: projects they will design, equipment they will use, and machines they will have to maintain. But also evident are what might be called organizational matters, including budgets, personnel decisions, and reorganizations. Two-thirds of the engineers interviewed (and over three-quarters of the *ingénieurs*) had subordinates for whom they had some responsibility; many had budgets or relied on departmental budgets. Almost all of them had developed ways to deal with and work through the current organizational structure. Dependent on these human, financial and organizational resources to do their own jobs effectively, they felt that they should be included, or at least consulted, in the decision-making about them, As with technical matters, this claimed right to participate is justified largely on grounds of efficiency. However, to the extent that such exclusion results in inadequate resources for their carrying out of assignments, it also unfairly disadvantages them in the competition for promotions.

To be sure, not all the engineers expressed their demands for more participation in the qualified terms used by those quoted above. A few called for more consultation in general or on "many things," including the general policies of the company or "at all levels." Several were angry about the level of consultation that did exist.

With regard to hierarchical relations here, many things are incomprehensible. There is no dialogue at all. There is just a monologue coming from the top down. [a TELECO *ingénieur*]

But only one engineer, a CFDT union activist at TELECO, called for more direct participation in all decision-making. And even he defended his stance by arguing that "decisions would be better because there would be more open criticisms."

It is also true that with respect to reorganizations or changes in "social policies" (vacation procedures, housing benefits, etc.) that may disrupt their private lives, there was strong feeling that action without consultation represents an abuse of power. This was especially so at TELECO, where paternalistic norms were weak. Employees there were more sensitive to any violation of the jurisdictional boundaries of the authority of office and of the implicit social contract to limit bureaucratic management to the domain of production. There was considerable outrage in the Simulator Division at the unilateral and unexpected decision by top management to transfer one-third to one-half of the engineers to a new site on the other side of Paris. Some of these engineers had just purchased houses near the old site and faced having to sell them or commute a long way. Several of these engineers participated in a protest demonstration at TELECO's headquarters. It is important to note, however, that this one example of engineers challenging management's authority was occasioned by intrusions of that authority into their private rather than working lives, contrary to the expectations of all three theories under consideration.

In the Telephone Division too there had been several recent reorganizations, some of which involved transferring whole departments from one city to another. There may be a causal link between these frequent reorganizations and the rapidly changing technology and markets of the electronics industry, or even of science-based industries in general. In the case of TELECO, however, the direct causes of most of the reorganizations were actions by the French government, including decentralization laws that limited expansion within the Paris region and industrial policies to promote the international competitiveness of the electronics industry by such means as state-sponsored mergers.

In any case, it is important to recognize the distinctively French character of the consultation grievances in French industry. In their comparative analysis of matched French and German firms, Maurice et al. (1984: 261) stress the differences in the nature of work relations in the two countries. They point especially to:

> the lower degree of autonomy in work and to the lesser importance of technical aspects of production [in France] as compared with the company's administrative functions, be they job preparation functions, new investments, commercial, or most important, personnel functions. It would seem that many decisions are taken in France without consulting production managers, whereas in Germany a production engineer or even a foreman is consulted on such questions as to where to install a machine, the form of a new product, a new production method, or a new way to organize work.

Most importantly, Maurice et al. (1984: 246) attribute the French pattern to the weakness of vocational education in France, and the resulting need and freedom a French company has "to develop its work organization in such a way that subordinate tasks needing only short training on-the-job are quite separate from planning and organizational tasks." While this gives technical departments a greater voice in determining production tasks, it also yields a "dependence on an administrative management that is relatively autonomous in relation to technical management (involved in production)" (p. 248). Personnel managers rather than production managers (engineers) determine job descriptions and assignments, for only a central office can assure consistency across departments in the absence of society-wide occupational practices (recognized apprenticeships, etc.). "The development of technical and administrative control functions inherent in this system goes hand in hand with a preponderance of hierarchical relations over relations of cooperation" (p. 248).

This explanation of strained authority relations within French industrial organizations stands in marked contrast to those that invoke global cultural attitudes towards authority (Crozier, 1964), yet helps explain those very attitudes.

The limits of participation

As members of the middle strata of a hierarchically organized system, however, French engineers are in a delicate position, and are understandably cautious about calling for greater participation by employees in organization decision-making. Further evidence on this subject emerges from two other interview questions. The first appeared in a follow-up written questionnaire sent several months after the interviews. It listed twenty-six specific areas of decision-making and asked the engineers to indicate their present and desired levels of participation in each one. Although only 20 of 52 PAMPCO engineers and 24 of 65 TELECO engineers completely answered this question, the pattern of the responses is too uniform to discount. Table 6.1 compares the number of respondents in each company who wanted either to make decisions alone in a specified area or at least participate directly in them (column B) with the number of those among them who felt they were not already participating at that level (column A). The most striking aspect of these findings is the small number of engineers who say they want to participate directly in decision-making in any but the most personal areas of their work lives. Not a single one of the 44 respondents indicates a desire to participate even in the appointment of top managers (c), the nomination of his department (e) or factory (m) head, or the decisions about company growth (l). Of a possible 176 responses to these four questions at the next highest level of involvement, only five indicate a desire to have one's advice taken into serious consideration. In seventeen of the twenty-six areas, including the determination of department budgets (k) and the establishment of product quality norms (t), seven or fewer responses indicate a desire to be at least directly involved in the decision-making. Only with respect to such personal matters as work clothes and tools (p), transfers to other positions (r), the replacement of machines in the engineer's own sector (u), and the setting and carrying out of daily work schedules (z), do as many as twenty of the forty-four respondents express a desire to at least participate directly in the decision-making. As for policy decisions on such general company matters as investments, products, and the selection of managers, most respondents don't even want to be consulted (levels 3 and 4); they just want to be informed.

A second striking pattern of the findings in Table 6.1 is the low number of engineers who do not already enjoy the high levels (levels 5 and 6) of participation that they want. Moreover, while not evident in the data summary in Table 6.1, the modal degree of the discontent was mild, i.e. a difference of only one level between the actual and the desired levels of participation. This is not to argue that most engineers were content. Many expressed an unsatisfied desire for more opportunities to express their opinion, or at least to be informed before a decision was executed. Finally, it is noteworthy that the response patterns in the two companies are quite similar. In short, while most engineers do seek more participation in the management process, the character of those demands is far from radical. At both PAMPCO and TELECO, the demand is rather to be better informed in general, more consulted on matters about which the engineers feel personally concerned or particularly qualified, and treated with the respect and trust seen as due French *cadres*.

Table 6.1. *Participation level desires (B) and discontent (A)*

Column B: The number of engineers at PAMPCO/TELECO who wanted to be at least directly involved in the making of such decisions (levels 5 & 6; see "*B* [levels]" below). Column A: Number, among those indicated in Column B, who said their actual participation level was 4 or lower, implying discontent. The total number of respondents was 20 at PAMPCO, 24 at TELECO.

On the subject of decision-making in your company, I would like to specify the areas in which you currently participate and those in which you would like to participate. Columns A and B represent these two variables; six possible responses are available for each case in the list below.

A (levels)	*B* (levels)
1) I am not informed.	1) I am not concerned.
2) I am informed before the decision is made.	2) I would like to be informed before the decision is made.
3) I can make my opinion known.	3) I would like to express my opinion.
4) My opinion is taken into account.	4) I would like to have my opinion taken into account.
5) I am directly involved in making the decision.	5) I would like to be directly involved in making the decision.
6) I alone make the decision.	6) I would like to make the decision myself.

Please indicate with the appropriate number your two responses to each of the following decisions:

		A PAM/TEL	*B* PAM/TEL
a	Investments made at the company level.	1/2	1/4
b	The introduction of a new product on the market.	1/1	2/1
c	The selection of general managers.	0/0	0/0
d	The determination of workers' salaries.	2/2	3/4
e	The appointment of a department head.	0/0	0/0
f	The organization of work in your department or workshop.	2/3	5/3
g	Economic layoffs.	3/0	3/0
h	The determination of production objectives for each department.	1/1	3/1
i	The management of social activities.	0/1	0/1
j	The determination of vacation periods.	3/4	5/10
k	The determination of the annual budgets of different departments.	1/3	1/3
l	The rate of growth of the company.	0/0	0/0
m	The nomination of the plant director.	0/0	0/0
n	Hiring regulations for salaried workers.	0/2	1/3
o	Access to company training sessions.	2/4	2/6
p	Acquisition of your work equipment (tools, clothing).	3/8	8/14
q	Improvements in working conditions.	4/6	8/9
r	Transfers to other positions in the plant.	7/5	9/10
s	Determination of the salaries of the directors.	3/0	3/0
t	Determination of standards for product quality.	1/0	4/3
u	Replacement of used machines in your sector.	5/5	8/12
v	The hours of work.	1/7	5/8
w	Rules about health and safety.	0/1	2/5
x	The use of time and motion studies.	1/6	4/10
y	The setting of prices in the company cafeteria.	0/1	0/1
z	The organization and use of your time during the working day.	3/3	13/23

Adapted from F. Dupuy and D. Martin, *Jeux et Enjeu de la Participation*, pp. 169–170.

Attitudes towards *autogestion*

A second interview question addressed the extent to which radical perspectives underlie the demands for more participation. This was a general question on workers' self-management (*autogestion*), a subject much discussed by the political parties on the left and by the French press in the 1970s. Only 5 percent of the engineers expressed a belief in *autogestion* as an important value in itself. Another 6 percent thought *autogestion* was an interesting enough idea to justify trying some form of it. An additional 25 percent displayed sympathy for the ideal, but thought it would not work in practice, at least not in today's large companies. The remainder – 64 percent at each company – were simply against self-management.

Closer inspection of the data shows that there is much greater variation by administrative unit than by firm. Of the 13 pro-*autogestion* engineers (including two in sales), all three at PAMPCO were *ingénieurs diplomés* who worked in the factory, and eight of the ten at TELECO worked in the Simulator Division. Furthermore, six of the seven most radical and committed responses were from engineers at the Simulator Division. These distributions suggest that it is something specific to particular units that is associated with a pro-*autogestion* attitude. And what stands out about both those units – judging from the reports of the engineers – is the particularly autocratic structures of authority in them. The fact that this structure is more paternalistic at the PAMPCO factory may also explain the lower incidence there than at the Simulator Division. Located in the conservative Lorraine and hiring very few engineers, PAMPCO could and did recruit engineers who fit the traditional image of a loyal *cadre*. TELECO's Simulator Division, by contrast, was unable to hire all the engineers its expanding business needed, and was recruiting them in the Paris region.

Finally, the less paternalistic climate at TELECO, the stronger external market position of the engineers there, and their large number meant that the TELECO engineers were freer collectively to oppose top management (Tilly, 1978). Thus, there was a *cadre* chapter of the militantly radical CFDT labor union at TELECO's Simulator Division, something the engineers at PAMPCO considered unimaginable. Moreover, the national CFDT vigorously supported the idea of *autogestion*, and so engineers who joined a local chapter of it simply because it was the one militant *cadre* union in their place of work were selectively exposed to pro-*autogestion* publicity. I shall return to these issues in the discussion of the collective organization of PAMPCO and TELECO engineers in chapter 8.

Those who were against *autogestion* emphasized the dangers of inefficient decision-making – due to endless meetings – and bad decision-making – due to the incompetence of the personnel. A few, but only a few of the critics displayed concern about the possible politicization of decision-making, abuses by the unions, or domination by the best talkers. Few of these critics ridiculed the idea of *autogestion*. Rather they said it wouldn't *work* in a capitalist world, or in France, or in big companies. And by "wouldn't work" they meant that the organization would become less effective at producing marketable products. In complex, economic organizations someone had to take responsibility, make decisions, and have the authority to ensure their efficient implementation.

Thus, for the same reasons that consultation was viewed as desirable, *autogestion* was seen as impractical. As one critic of *autogestion* put it: "*Autogestion* won't work in big companies, and more participation [systematic consultation] would suffice to involve the lower ranks, thus motivating better performances and increasing efficiency." The defenders of *autogestion* also emphasized its potential to "generate new energy and force" and to get the workers to "work more because they would be more interested." Moreover, they acknowledged concerns about its practicality, and suggested that implementation would have to be gradual, accompanied perhaps by new forms of education. In short, both bureaucratic and democratic forms of management were judged on instrumental grounds – not as ends in themselves but as better or worse means for achieving other ends. And those other ends are not the liberation of human beings from alienating forms of work, but the efficient administration of industrial enterprises.

The uses of authority

The discussion thus far has identified the nature of these engineers' attitudes towards participation and *autogestion*, but has done little to explain them. One explanation, of course, was implicit in the discussion of autonomy: freedom from direct supervision in daily work spares engineers from frequent negative experiences with authority. Moreover, most of them like their work. Consequently, they feel little need to curtail managerial authority. There is, however, a second side of the engineers' experience with authority that strengthens their favorable disposition towards it. Hints at it appeared in the expressions of concern for responsible leadership. This section further examines the positive side of bureaucratic authority for engineers.

Each engineer was asked: "In the kind of work that you do, do you think it's necessary to have a supervisor [*chef*]?" Perhaps unsurprising, in view of the preceding discussion, 73 percent replied that a *chef* is "important," "necessary," or "indispensable"; 21 percent that a *chef* is useful, but not critical, and only 6 percent that a *chef* is of little or no use. Several added that while a *chef* is necessary, there are too many of them and too many hierarchical levels at TELECO. The differences between PAMPCO and TELECO are tiny, and the only noteworthy variation among the respondents is a slight tendency for *techniciens* to see less need for a *chef*.

More interesting for the present purpose are the responses given to the probes about why a supervisor is important and exactly what role(s) he performs. Table 6.2 displays the categories into which the responses were coded and the percentage in each firm that gave this response.

To handle the many multiple responses to this question, Table 6.2 combines the primary responses by 105 engineers with the secondary responses of the 85 among them who indicated at least two important roles a *chef* plays. Inspection of the "combined" column in Table 6.2 reveals that few engineers regard a supervisor as needed for either the technical assistance or the control he may provide. Rather, these engineers value a supervisor for the various coordinating roles he performs. The table distinguishes three specific types of coordination: (1) non-authoritative internal coordination; (2) authoritative internal coordination, and (3) coordination with the rest of the company, which is seen as serving an important need. Both authoritative and non-authoritative internal coordination refer to the organization and integration within a department or team of the more specialized work of individual engineers. Some thought such coordination could be accomplished by a coordinator at the same hierarchical level as the other engineers, but most felt the coordinator needed sufficient authority over the others to resolve differences of opinion and make binding decisions. In view of the expectations of "legitimation crisis" theory, the following interview is drawn entirely from TELECO engineers.

The *chef* should be a coordinator and advisor. He is necessary because the *chef* has an overview (*vue de l'ensemble*) of the entire department that we in our separate sections lack.

A project *chef* is indispensable for coordinating the whole business. It's necessary to have someone who coordinates the ideas of each, to make a plan, and once it is made to follow and coordinate the progress of each sub-group.

As long as my work is part of a larger complex whole and has consequences for it, I need a *chef* to orient my work. We have autonomy and initiative, but need coordination and decision-making.

There are always conflicts to be resolved. It was a real problem in one factory where the boss wouldn't make decisions. Efficiency requires objectives, which can only come from and be realized with the help of a director.

It's necessary to have someone who makes a decision and takes responsibility. Must avoid awful, day-long conversations which don't end in a decision, especially on important projects.

In this kind of work you need organizers and arbitrators. And they must have authority. Otherwise someone else will assume it, which goes to prove the need.

In general, these comments show the felt dependence of engineers on

Table 6.2. *Role of supervisors, by firm*

		PAMPCO %	TELECO %	Combined %
1	Technical advice	10	11	10.5
2	Non-authoritative internal coordination	11	18	15
3	Authoritative internal coordination: work organization and final decision-making	48	37	42
4	Coordination with rest of company	18	29	24
5	Control: maintaining standards, evaluating personnel, discipline	13	5	8
		100	100	99.5
Total number of responses		88	102	190
Total number of respondents		49	56	105

others for the accomplishment of their own tasks and the resulting desire for effective leadership to organize, coordinate, and expedite group activities. Most "professionals" are less dependent on others for the successful performance of their work tasks than are engineers. Teachers, doctors, and lawyers tend to work in what Lortie (1975: 13–16) calls a "'cellular' pattern" of relative isolation from, and low task interdependence with, colleagues. Engineers, by contrast, are specialists who rely on both other engineers and many non-engineers to render their own efforts effective (Larson, 1977: 25–31). In short, the engineer views well-exercised authority as facilitating his work, both in the immediate sense and by means of its contribution to organizational success. Even the more critical and radical TELECO engineers made this point.

I am for decentralization. Each individual ought to be free to make decisions at his level but there also needs to be a hierarchy which integrates the whole.

In the final analysis, it's necessary that someone make a decision. When I speak of authority, I mean that the authority should stem from the fact that this person is recognized as being competent and not represent authority conferred from above. It is a bit utopian, but it is imaginable.

The engineer from whom the last quote was taken was an organizer and activist for the CFDT, a union that officially advocated *autogestion*. Yet, even he finds the idea "a bit utopian."

One of the reasons for concern about decision-making is the understandable fear of responsibility. In the world of industrial engineering, design decisions can have enormous consequences for entire projects and product lines. "Bad decisions" may result in cost over-runs, cancelled contracts, product recalls, consumer suits, or simply negligible sales or profits. Several TELECO engineers mentioned their concerns about having too much responsibility for decisions beyond their competence:

I had too much technical and financial responsibility three years ago. It was scary to make such decisions alone. [in answer to earlier question on autonomy]

I'm for a consultative hierarchy. I avoid making decisions alone; do so only as a last resort.

In addition to departmental coordination and decision-making, many engineers said that a major role of a *chef* is to represent his department in its relations with other departments and higher management. In this role a good *chef* could help his subordinates by protecting the entire department's interests in various intra-organizational struggles for budget and staff increases, desirable projects and assignments, promotions, favorable resolutions of inter-departmental conflicts, and numerous small favors. And a good *chef* in this case is one who is influential in the larger company and knows how to "work the system," in short, an effective organization man.

You need a *chef* to act as an intermediary with higher management. Mine has sponsored me within the company a bit, for he knows people. His weight helps in these relations with the rest of the firm. [a TELECO *ingénieur*]

We must depend on someone to manage our relations with other departments. And even better a superior who can deal strongly with other departments. [a TELECO *ingénieur*]

A *chef* who contents himself with supervising my work is not worth the trouble, but someone who deals effectively with other departments and provides liaison with the higher levels of management, that makes sense. [a TELECO *ingénieur*]

What about the differences between the PAMPCO and TELECO distributions, concerning the value of a *chef*? According to Table 6.2, the TELECO engineers give greater emphasis to the value of a *chef*'s non-authoritative coordination and departmental representation roles, while the PAMPCO engineers show more concern for the authoritative coordinating and control roles. These differences appear to reflect the slightly different authority situations of traditional and high technology engineers in France, and warrant closer examination.

Thus far, this analysis has ignored the position of engineers as supervisors themselves. It would be a serious mistake, however, to overlook the fact that as *cadres*, all *ingénieurs* have a stake in the current authority system. Seventy-eight percent of them have it in the immediate sense of currently exercising some supervisory responsibilities; all of the *ingénieurs diplomés* have it in the longer-run sense of being very likely to attain at least the position of *chef de service* sooner or later; and all *cadres* have it in the collective sense of their official status as managers. Even among the *techniciens supérieurs*, 40 percent report having one or more subordinates at present. This figure is higher than the 29 percent for the sample of ninety-three *techniciens supérieurs* studied by Fossati and Said (1983), but 42 percent of their sample was under the age of twenty-five. Quite a few of these *techniciens supérieurs* will end their careers as *ingénieurs autodidactes* (i.e. *cadres*), where the percentage with subordinates in the PAMPCO–TELECO sample is 72 percent.

But if, as *cadres*, all French *ingénieurs* have a stake in bureaucratic authority, it is as active supervisors or workers that they most strongly experience the need for and value of authority over subordinates. And while at both companies roughly similar proportions of *cadres* have subordinates, it is at PAMPCO that more engineers deal with more workers more of the time. As a mass-production metal-working company, PAMPCO employs many manual workers and few engineers. Production is the central function, engineers supervise production, and they do so in conditions of constant pressure for increasing output, reducing costs and maintaining quality. Consequently authority relations are intense.

The fact that workers' unions have gained representation at the factory level since 1968 complicates the authority situation of the production engineers, for union representatives now have local strength and direct access to local top management. In the eyes of several PAMPCO engineers, management has grown too fearful of the unions and consequently often fails to back up the contested commands of supervisors. Thus, while 73 percent of research and design engineers felt it is easier to command subordinates today than in the past, only 28 percent of production engineers agreed. Production engineers also stand out for their hostility to workers' unions, and the relatively larger number and greater centrality of production engineers at PAMPCO probably explains why 67 percent of PAMPCO engineers but only 31 percent of TELECO engineers express hostility towards workers' unions.[6]

In short, with respect to their own positions of authority over subordinates, the situation of PAMPCO engineers is considerably more problematic than that of TELECO engineers, and it is this fact that probably explains, at least partially, the greater emphasis at PAMPCO upon the control and authoritative coordination uses of authority.

While the engineers at PAMPCO are more concerned about supervision of their subordinates, the engineers at TELECO are more concerned about effective leadership by top management. The reason is that the telecommunications and electronics industries in France were undergoing enormous technological and organizational upheavals, and seemed likely to continue doing so. Thus, it is not surprising that several Telephone Division engineers expressed concern about the capacity of top management to successfully steer TELECO through the challenges ahead. As one TELECO *ingénieur diplomé* explained in response to the question on participation:

I'm especially concerned about the ongoing restructuring of the company, about problems of reconverting people and whole factories . . . Top management seems to be moving slowly, avoiding necessary decisions, and making contradictory ones. What I want is good management and leadership more than to participate myself.

These differences in the authority situations and attitudes of the PAMPCO and TELECO engineers stand out in the responses to a question about morale problems at each company. While a larger proportion of PAMPCO engineers mentioned problems of authority relations and organizational tension (38 percent vs. 22 percent at TELECO), a larger proportion of TELECO engineers expressed concerns about management's capacity to lead (16 percent vs. 6 percent at PAMPCO). It would be foolish to make a great deal of these differences, but they do suggest that if science-based industries are changing the authority situations and attitudes of French engineers, it is hardly in ways that threaten the legitimacy of management.

If French engineers accept bureaucratic authority in principle, what then accounts for their frequent and often harsh criticisms of it in practice? The answer appears to be their perceptions of individual supervisors as authoritarian, secretive, and in general "bad *chefs.*" It is not the system, but rather Monsieur X or Y.[7] The following comments, taken from replies to the question about the desirability of greater responsibility and decision-making freedom, are illustrative.

Yes, it's too centralized here. It was the opposite at XYZ [a different PAMPCO factory]. Here Monsieur X [the head of the whole Division] decides everything. [PAMPCO]

Yes, Monsieur Y [his supervisor's supervisor] tends to be traditional in his conception of authority, but he's beginning to loosen up. [TELECO]

This same emphasis on the individual occupant of a supervisory position appears also in some of the more favorable responses.

Here I'm relatively free. I can spend money as I see fit, within the limits of my annual budget. I can exercise initiative. It was worse at the Research Center . . . Of course, I still clear everything with Monsieur X [his *chef*]; he makes the final decisions. It's only on details that I act independently. But that's natural. I accept the hierarchy and like my current *chef*; he is accessible, consults with me, and listens to me. [PAMPCO]

I have no budget of my own; my *chef* has it. To order any materials for mock-ups I must see him. But it's not a problem, and I have a lot of autonomy here. I am free to organize my work as I wish. It was quite different three years ago with a *chef* who supervised the *ingénieurs* very closely. I appreciate Monsieur Y's [his current *chef*] trust of me. Before we were treated like *techniciens*, but one reason was that the *chef* was an *ingénieur maison* who didn't understand that we *ingénieurs diplomés* were not *techniciens* and could take responsibility. We organized and resisted and eventually forced a reorganization of the group, with the *techniciens* being assigned to *ingénieurs* rather than all being equal. [TELECO]

This last comment is particularly revealing. First, trusted workers take pride in being trustworthy. Second, not only are there good and bad supervisors, but bad supervisors may be resisted. Such collective resistance was also evident in the demonstrations at TELECO headquarters against the unilateral decision to transfer several Simulator Division departments to the far side of Paris. In both cases we see angry engineers taking fairly militant action in behalf of demands for more autonomy and participation. But these are quite moderate demands for more respect, trust, and consultation of *cadres*, not radical challenges to bureaucratic authority or occupational inequalities, and they are found at both companies. Moreover, while the authority situations and the attitudes towards workers' unions and the uses of authority do differ at PAMPCO and TELECO, the major reason for the greater militancy at TELECO is not a more radical ideology (although see chapter 8 for some differences) but rather a greater capacity and more opportunities for collective action. As argued earlier, group protest is easier in companies where there are many lower level engineers, where they are less isolated from one another and more isolated from their bureaucratic superiors than at PAMPCO, and where a less paternalistic climate and better external market situation than at PAMPCO make collective action less risky.

To summarize, in both their status as subordinates and their status as managers, French engineers, in science-based as well as traditional industries, find bureaucratic authority useful for achieving their own work tasks and career goals. Dependent on numerous other specialists and workers, they especially value the coordinating and representational roles that supervisors can perform. They are sensitive to what they see as abuses of authority, and may react strongly where possible, but they tend to blame such abuses on individual "bad *chefs*" or at worst, the managerial style of particular companies. Occasionally, they blame French culture in general, but still not the principle of bureaucratic authority. Enjoying considerable freedom from direct supervision and experiencing authority as necessary and largely benign, they are understandably indisposed to question the legitimacy of bureaucratic management in general.

Attitudes towards profit

I have argued that engineers accept bureaucratic authority not because of its legality, but because they view it as a rational and efficient mode of organizing production and because they have a stake in it. However, in capitalist societies, the *raison d'être* of industrial bureaucracies is not the rationalization and maximization of production, but rather the maximization of profits for their owners. The question then arises: do engineers also accept bureaucratic authority in the service of capital. Thornstein Veblen, Serge Mallet and André Gorz maintain that "by training, and perhaps also by native bent," (Veblen, 1965: 74) engineers are naturally critical of the subordination of research and production to the logic of capitalist profit-seeking. Daniel Bell, too, although from a quite different perspective, expects engineers, especially those in the science-based industries, increasingly to reject the prevailing business ethos of the past century as they assimilate the professional norms (service, social responsibility) of a technical intelligentsia in a post-industrial society.

In order to address this issue, the interviews included the following question: "Are you ever unable to do as good a technical job as you would like because of business considerations – time, costs, etc.?" Table 6.3 identifies the coding categories and displays the distributions of responses by company. The vast majority of engineers began their answer with "Yes, it happens to me very often," or a similarly affirmative response, but then went on to explain the inevitability of some such compromises. Only two engineers, one in each firm, blamed excessive concern with business considerations for technical compromises, and their criticisms were mild and ambivalent. For example:

Yes, it's true. The company limits us to a certain range of electronic components which it stocks, and we can't choose others that would sometimes be better. I'm used to it now, but it was frustrating at first because you have to modify your design accordingly and settle for less satisfactory results. The company shouldn't impose such compromises with quality just to reduce costs. [a TELECO *technicien supérieure*]

Moreover, this respondent refused to blame capitalism or the profit system, even after repeated probes. Rather, he blamed management. Many other engineers also blamed management for at least some technical compromises, but these criticisms were more clearly aimed at false economies, unnecessary rushes to please a client, and other examples of what they saw as bad management.

Sometimes they judge machines according to price alone, without sufficiently considering what we could earn eventually. [a PAMPCO *ingénieur*]

Table 6.3. *Attitudes toward technical compromises*

		PAMPCO %	TELECO %	Combined %
I	Critical of profit-motivated compromises	2	2	2
2	Critical of some compromises but blames management: short-sighted cost-cutting, sales staff	20	18	19
3	Both [2] and [4]	6	15	11
4	Accepts compromise as part of job: company must sell, make profit, meet contracts, etc.	45	47	46
5	Sees no conflict or compromises	25	13	18
6	Other, don't know	2	5	4
		100	100	100
	Total number of respondents	49	60	109

Yes, often. In general it makes me mad because when something is badly made, it works badly, and it's necessary to repair it, and that's more expensive. [*a TELECO* ingénieur]

Although these responses blame bad management rather than business values, they do not demonstrate a clear acceptance of business considerations and so were coded into row 2. However, 57 percent of the engineers ("combined" column, rows 3 and 4 together) do accept making technical compromises imposed for business reasons, whether or not they also blame bad management for some unnecessary compromises (row 3). Some of them even embraced the challenge of finding commercially viable solutions.

Sometimes we abandon solutions that we know are better (for financial reasons). Because what we're seeking essentially . . . is the best quality–price relationship . . . That's part of our craft (*métier*) . . . a rule of the game. It's not a source of dissatisfaction. [a PAMPCO *ingénieur*]

Yes, constantly. But constraints are necessary, part of an engineer's job (*métier*) is to achieve the objective quickly and cheaply. Perfection doesn't exist, and finding the best compromise is what makes the work interesting. Those who feel compromised are professors, not engineers – who are practical men – and should work in pure research. [a PAMPCO research *chef*]

Yes, costs, always costs. But if there wasn't any consumer, there would be no production, no jobs. It's necessary to make a product that is good, but as cheaply as
possible. It's a system that means the whole end result is positive, productive. [a
PAMPCO *technicien supérieur*]

Absolutely. It's often frustrating. But these are the imperatives that exist . . . It's the
law of the market and of contract penalties. [a TELECO *chef*]

Yes. All the time. Unrealistic deadlines. Or we try to do something the client wants,
even though we don't know how very well. But that's our profession (*métier*). I
should respect the rights of those who provide the work. I am the orderly (*tampon*)
whose job is to realise the client's idea. [a TELECO *ingénieur*]

We could have made the X system better, but it would have been more expensive.
It's distressing but necessary. Those are the rules of the game. I accept the market
system. My role is to make things at the lowest possible price. [a TELECO
ingénieur]

Clearly, these engineers do experience a conflict between the demands of
profitable business and their own desires to produce technically impressive
products. However, the experience is not one of a conflict of values, but
rather of a natural tension between personal inclinations and the realities of
professional engineering. The major realities for engineers are the market
and the law of exchange. Business enterprises are viewed as largely neutral
agents reacting reasonably to the imperatives of survival in the market. The
engineers' view of business does not celebrate capitalism. None of the comments mentions the rights of private property, the functions of profits as
incentives to risky investment, or the deserved rewards for creative entrepreneurship. Rather, they reflect a conception of capitalism as neither
heroic nor exploitative, but as natural and necessary. Thus, the frustrations
notwithstanding, business-mandated cost and time pressures appear to be
accepted by the majority of respondents as one more property of the natural
order which engineers must take into consideration in making their
calculations.[8]

   Table 6.3 also shows 18 percent of the engineers reporting that they have
not had to accept technical compromises for reasons of business considerations. Half of these are *techniciens* whose more routine work (e.g. in
research or test labs) insulates them from the pressures of the market.
Having little say in the choice of technical solutions, *techniciens* do not have
to agonize over the best quality–price compromises. Such naivete may also
explain why, when they do feel thwarted technically, *techniciens supérieurs*
are far more likely than other engineers, to point to bad management rather
than the realities of the market. It should be added, however, that by training, non-*cadre* status, and the logic of their career prospects, *techniciens*
would seem less likely anyway to identify with management and its problems. The influence of technical training is of concern, because Veblen,

Gorz, and to a certain extent Bell emphasize it as a source of anti-business values. Ironically, in contemporary France at least, the more a student progresses in technical schooling, from the technical *lycée* to the *grande école*, the more he is pulled out of the separate occupational culture of the *technicien* and, through courses and anticipatory socialization, into the business culture of *cadres*.

Conclusions

This chapter has argued that although there is some dissatisfaction among French engineers over their relative exclusion from some decision-making, there is little disposition in either traditional or science-based industry to question the legitimacy of bureaucratic authority or business values. It has also suggested some reasons for this being the case, pointing in particular to the freedom from direct supervision and to the positive uses of supervisory authority. These are reasons that are inherent in engineering work, and correspond closely to the findings of Whalley (1986) and Zussman (1985) in Britain and the United States. However, French industry is distinctive in its traditions and practices of defining engineers as managers, and in the degree to which the organization decides about the structure of positions and the assignment of personnel to them. This affects recruitment to the occupation, and gives French engineers a personal stake in the legitimacy of bureaucratic authority.

Yet, the reality of the authority situation of many of them does not live up to their expectations. These expectations are shaped by an image of *Monsieur l'ingénieur* as an officer in the industrial army and a leader of men. This image is rooted in the reality of French military engineers in the eighteenth century. It was enhanced by the decision of the first non-state engineering school, the Ecole Centrale, to model itself after the prestigious Ecole Polytechnique. It received affirmation among industrial engineers towards the end of the nineteenth century with the emergence of paternalistic strategies that shifted supervisory authority from foremen to engineers.

It is the period between World War I and World War II, however, that sees the flowering of a well-formulated ideology of the engineer as commander. This timing may reflect the fact that it was only after World War I that the state engineers, who began moving into the private sector in the 1890s, started to reach the top of companies that were increasingly free of external control. As Levy-Leboyer (1980: 133) points out, these executives had been trained to manage non-profit bureaucratic organizations such as the army, state railway, and state-owned mines, industries characterized by stable technology and large labor forces. Thus, "their primary concern was with hierarchy, leadership, the division of authority, and work discipline."

Their ascendancy in private industry coincided with the after-shocks of the Russian Revolution, which generated new anxieties about labor discipline.

Whatever the reasons, there appeared in the 1930s and 1940s a series of books and articles emphasizing and glamorizing the leadership responsibilities of the engineer. The best example of this literature is the influential book by Georges Lamirand, first published in 1932. Boltanski (1987: 81) summarizes it well:

> The engineer, writes Lamirand, is a leader: he must "serve and command" (p. 40), "gain the sympathy of his men by his manner" (p. 50), "look his men directly in the eye" (p. 55), and "impress them . . . with the force of his mind and will" (p. 68). He must "give the impression of physical superiority" (p. 55), which can be achieved through gymnastics and sports, and in short he must possess the virile qualities that make an officer: frankness, a firm sense of reality, courage, tenacity, and dedication to his work . . . Finally, the engineer must "know how to punish." (*pp. 56–57*)

To be sure, this *Monsieur l'ingénieur* is largely mythical by the time Lamirand is writing, and represents more the wishes of employers than reality. As Kolboom (1982: 72) explains, "One of the consequences of the second industrialization and the rise of the large, bureaucratic corporation is the emergence of the salaried technician and the disassociation of the symbiosis between *patron* and *ingénieur*, who was a sort of assistant *patron* in the past." Nevertheless, the myth is powerful, and it is little wonder that many French engineers in today's high technology companies feel that they are not really engineers, that they are only technicians. Commanding few if any subordinates, they cannot view themselves as leaders of men on the industrial battle field. And while the army of troops has evaporated from beneath them, the hierarchy of superiors over them has extended beyond sight. Between the rapid post-war growth in the size of French firms, the new emphasis on marketing rather than production, and the disproportionate expansion of engineering staffs in the science-based sector, many French engineers feel deprived of their traditional place in the industrial hierarchy. In the high technology companies in particular, they no longer command, and they are no longer consulted.[9]

Some observers interpret these developments as a kind of proletarianization. But proletarianization refers to a decline of intrinsic and extrinsic rewards due to successful management efforts to deskill work and exercise tighter control over the labor process. There is no evidence in these findings of closer supervision, either direct or indirect, in the science-based company. Indeed, the opposite seems to be the case, perhaps because the growth in the numbers of *ingénieurs* per *chef* and the historical decline in French paternalism has made *ingénieurs* in science-based companies less visible and less subject to the arbitrary discipline of a stern patriarch.

From time to time, there have been outbursts of protest that express both the frustrations and collective capacities of engineers in the high technology sector. Such militancy misled Mallet in two ways. Not only do the demands for more participation represent far less radical challenges to the system than he imagined, but they also represent tensions characteristic of French industry rather than science-based industry everywhere. While Mallet over-generalized from the French situation, theorists of professionalization seem to have ignored it. French engineers are not challenging bureaucratic authority as engineers in the name of occupational rights; rather, acting as *cadres*, they are demanding a greater share of bureaucratic authority in the name of traditional *cadre* privileges.

The way that they obtain such authority, however, is not through collective action, as they well know, but through individual advancement to higher positions. It is to the career structure of technical work and changes in it that we must turn to gain further insight into the position and attitudes of the French engineer.

7

Labor markets and career experiences

Everyone's career is essentially determined once and for all by the diplomas, titles, or grades obtained at the outset of his career. Raymond Aron, *The Illusive Revolution* (1968: 46)

Theories of proletarianization, a new working class, and professionalization focus on work content and work situation. That is, in attempting to explain or predict the identities and ideologies of technical workers, they refer to the skills these workers use, the jobs they do, and the location of those jobs in the divisions of labor and authority. The preceding chapters have followed a similar course, examining in turn the knowledge engineers use, the organization of their work, and their authority situations. In the process, they have criticized the theories for neglecting the market situations of technical workers. This chapter and the next attempt to honor the implicit promise of those criticisms by investigating more systematically the character and significance of the labor market for French engineers. They give special attention to engineering careers and the specific way they are organized in France.

One reason careers are important is that both past work experiences and expectations for the future affect individuals' responses to their present jobs. Burchell and Rubery (1987) show that job satisfaction is considerably higher when individuals view their current job as the best job they have ever had. Generations of ambitious youth bear witness to the reduced significance of unpleasant jobs when they are seen as a stepping stone to a better future. Wherever careers are visibly structured, the understanding of any single job, by both occupant and employer, is affected by that job's position in a career ladder.

In addition to affecting job satisfaction, careers influence occupational identities and solidarities. Because military and medical careers lead all members of those professions through a similar and well ordered series of steps, they engender a sense of membership in a single professional community. In other cases, however, the career experiences and prospects of experts in similar jobs may vary widely, breaking down solidarity among them and generating different interests and attitudes.

One obvious way in which careers may produce different interests and sentiments is by leading to different positions in the stratification system. Stratification research, whether employing status attainment or human capital models, has long focused on the individual determinants of career outcomes. More recently, however, sociologists and labor economists have recognized that "rewards are attached to organizational positions and that organizations differ systematically in their personnel practices and reward systems" (Baron, 1984: 37–38). Baron emphasizes two ways in which organizations shape careers: "First, the division of labor among jobs and organizations generates a distribution of opportunities and rewards that often antedates . . . the hiring of people to fill those jobs. Second, organizational procedures for matching workers to those jobs affect the distribution of rewards and opportunities within and across firms, and thus influence the likelihood of career success."

This chapter on the labor market situation of French engineers addresses these and related matters. It begins with a brief discussion of engineering careers in the past and of historical developments affecting them. The following two sections focus on the structures of opportunity today, by examining in turn the typical promotion ladders within French firms, and the significance of credentials and seniority in career mobility. The discussion then turns to job security and salaries. The following section looks at recruitment and socialization to the occupation of engineering in France and to specific firms. The chapter ends with a detailed analysis of the attitudes of the PAMPCO and TELECO engineers towards their job security, salaries, and careers.

Recent developments in French engineering careers

For the past thirty years, many French engineers have enjoyed careers involving considerable upward mobility. This has been facilitated by the rapid growth of the French economy in general and the engineering sector in particular. Yet, this general growth at all levels of the occupational hierarchy has obscured what appears to be a disproportionate growth in the numbers of lower ranking engineers. The FASFID surveys are inconclusive as evidence on this point, but they do suggest such a shift. Between 1967 and 1979, the percentage of *ingénieurs diplomés* working as ordinary *ingénieurs* rose from 40 to 43, and the percentage working as *chefs de service* rose from 24 to 28; the percentages working in higher positions showed a corresponding decline (FASFID, 1968: 34; 1980: 45). Moreover, such a broadening of the base of the engineering hierarchy is to be expected if the distribution of hierarchical positions is more skewed towards the lower levels in the rising, high technology industries than in the traditional industries. In fact, it seems to be. For example, while only 31 percent of *ingénieurs diplomés* in the mining, forge and foundry sector were ordinary *ingénieurs* in 1977, in the electrical machinery and electronics industries 50 percent were (FASFID, 1977: 66). The only other industrial groups to reach 50 percent were aeronautics–armaments (54 percent) and energy (50 percent). By contrast, the figure for textiles–shoes was 17 percent. Finally, according to the UIMM (1977: 22) survey of metal industries (including automobiles, electronics and aeronautics), the proportion of *cadre* positions defined as top management (*direction*) has declined substantially, from 10.7 percent in 1956 to 5.9 percent in 1975. The compensating growth has been in sales and administration, not in the technical areas.

From a historical perspective, some engineering careers seem to be peaking at lower levels. According to C. R. Day's study of 800 *gadzarts* during the years 1825–90, when Arts et Métiers was only a technical high school, 70 percent began their careers as skilled manual workers or draftsmen, but by the age of retirement, "two-thirds [of those in industry] had become at least upper-middle supervisors, engineers, and owners" (1978: 453). Sixty-one percent of those in private industry became proprietors, and of these, 79 percent became what Day calls "substantial owners." For engineers lacking a higher diploma, such mobility is rare today.

Yet, it would be premature to conclude that the career prospects of *ingénieurs diplomés* are declining. Examinations of positions – of the proportions of all *ingénieurs* who are ordinary *ingénieurs* and/or *chefs* – overlook the impact of people on positions and the rate of mobility through positions in expanding organizations. It is entirely possible that employers adjust career hierarchies to meet their current needs for motivating employees. Firms recognize the importance of rewarding trusted workers with promotions, often creating "dual career ladders" to accommodate them. When the engineering staff is composed largely of recent recruits fresh out of school, such promotions are prospective, but as that engineering staff ages, firms are obliged to fulfill the earlier promises of advancement for a larger proportion of engineers. The proportion of *ingénieurs diplomés* under age 35 increased from 36 percent in 1968 to 42 percent in 1980.[1] Only after the engineering occupation begins to age again will it be possible to determine whether the recent shifts in the ratios of higher-to-lower level positions represent a temporary adjustment to changes in the age distribution of engineers, or a more permanent broadening of the base of the engineering industry.

However, even if engineering careers are peaking at lower levels than in the past, it is far from obvious how such a development should be interpreted. True, those without degrees from engineering schools are no longer likely to rise far in the managerial hierarchy, but a larger proportion of the industrial work force now has engineering diplomas, and practically none of these *ingénieurs diplomés* ever works as a skilled worker or draftsman, much less ends his career at that level. Moreover, until 1934, *ingénieurs diplomés* often suffered from their lack of any official status; in the automobile industry, for example, they were often listed in company records as *dessinateurs*, etc., and lacked a contract (Fridenson, 1985). Finally, although the average *ingénieur diplomé* may not rise as far as he once did, the graduates of the most illustrious engineering schools continue to enjoy rapid advancement to top management positions, albeit with increasing competition from the growing ranks of business school graduates.

The recent rise of business schools and their graduates in France is relevant to this study's larger arguments in two ways, one concerning the causes of this phenomenon, the other the consequences for engineers. Taking the latter first, recall that after World War I the state-trained engineers, who had abandoned the public sector in the 1890s because it was oversaturated and underpaid, had begun achieving top management positions in the private sector. Business historian Maurice Levy-Leboyer (1980: 134) notes that "the ideal of that generation, with its bureaucratic and engineering interests, was not so much the development of the market . . . but the establishment of functional organizations, logically built, where a staff made up of highly trained engineers would set the rules."

The Depression undermined the premises of this production rationalizing approach to business, but it was only after World War II – decades later than in other advanced societies – that the multi-divisional, marketing-oriented enterprise became common among large French companies. Before World War II, French managers received almost no training in economics or industrial psychology, and there were virtually no higher schools of business administration. The 1950s and 1960s, however, witnessed a wave of fascination with American management practices, the rapid rise of business schools in both numbers and prestige, and the emergence of a new career route into middle and higher management (Boltanski, 1987: 117–141). The post-war period has also seen a major increase in the percentage of company directors who are graduates of Sciences Politiques and the Ecole d'Administration National.[2] Even within the metal-working industries, the percentage of *ingénieurs diplomés* who occupied top management posts declined from 14 percent in 1956 to 7.8 percent in 1975. This is not surprising, given the disproportionate growth of *petites écoles d'ingénieur*. Yet, the figures on graduates of the elite engineering and business schools do suggest a weakening in the capacity of engineers to move into top management.

Competition from graduates of non-engineering schools may be reducing the chances of engineers, especially graduates of middle level schools, to achieve the kinds of managerial positions their counterparts did twenty-five years ago. To call this proletarianization, however, seems questionable.

Engineering work in itself is not thereby deskilled, and a shift in favored routes to higher management implies no overall decline in the numbers or proportions of managerial positions. Moreover, in France, where *grande école* diplomas qualify graduates for certain organizational levels rather than specific occupations, and where education is *polyvalent* and training is done on the job, many engineers are able to take advantage of the new career opportunities in finance, personnel management, and especially sales.

This is not the place to analyze the causes of the rapid rise of business schools and specialists in France.[3] However, it is important to note that the sudden adoption by the French of American-style management training programs seems to have been the outcome of two historical developments that are unrelated to any logic of industrialism or of capitalism. The first is the aggressive promotion of American management practices by the United States after World War II. The major objective of the Americans was to reduce the appeal of communism to French workers by "modernizing" the economy and labor–management relations. The major vehicle for inducing the desired transformation was the Marshall Plan. "Above all," said the American advisors sent to France under the Plan, "French firms should find devoted and efficient 'middle management' personnel and first of all learn how to train them ... Training of new managers and the creation of business administration schools . . . thus constituted 'fundamental duties.'" (Boltanski, 1983: 377).

The second historical development concerns the French response to this pressure from the Americans. Boltanski is quite correct when he writes that:

the mediations through which the [American productivity] missions took effect cannot be grasped without abandoning the mechanistic diffusionist models which they invoke, a universal determinism of an economic and technical nature . . . or "imperialist violence." Thus, it is necessary to analyze, on the one hand, the collision of the American model and the older French image, and, on the other hand, the struggle within the French bourgeoisie concerning the introduction and diffusion of the American model. (*Boltanski, 1983: 378*)

The latter raises issues specific to recent French industry, including the roles of various groups under the Vichy regime and in the war-time resistance to the German occupation. It is hard to imagine that the American models of management training would have been nearly so influential had they not suited the political interests of the ascendant groups within French business circles after the war.

The structure of positions

The developments discussed above represent society-wide aspects of the structure of opportunity for French engineers. In other ways too, French engineers face similar situations. French organizations tend to have more hierarchical levels, more white-collar workers, and larger wage differentials than their German counterparts in the same industry (Maurice et al., 1977). French companies offer certain state or collective bargaining mandated privileges to certain categories of employees according to an "officially recognized system of classification" that is particular to France (Stark, 1986). For example, *cadres* are entitled to the benefits of a special supplementary pension fund. (Yet, individual employers decide who is a *cadre*). Finally, the occupational hierarchies of many large firms are organized according to a single model that officially governs the wide variety of companies in the private metal industries, including PAMPCO and TELECO. This model warrants closer inspection.

In France the *conventions collectives* – industry-wide management–union agreements – define three basic groups: workers, collaborators (office workers, foremen, and technicians) and *cadres*. Industrial sociologists tend to add a fourth, *cadres supérieurs* (largely exempt – *hors statut* – from the *conventions collectives*), and to break up the lower layers along the functional lines that define normal career paths. Table 7.1 shows this organizational ladder of opportunity as described by two industrial sociologists, Claude and Michelle Durand (1971: 14). The three digit numbers down the left side refer to a national salary scale, the Parodi Index.[4] Minimum wage jobs are rated 100. Of immediate interest is the difference in the situation of *techniciens supérieurs* and *ingénieurs débutants*. Although at similar levels of the Parodi Index, the *ingénieurs* – both *diplomés* and *nondiplomés* – are at the bottom of their ladder and will easily advance to the top of it if they are not so old as to run out of career time. By contrast, the *techniciens supérieurs* are near the top of their particular ladder, and must first surmount the *cadre* barrier before advancing. *Techniciens* who have the BTS or DUT degrees tend to be hired at the level of Coefficient 255, depending on the regional and industry-wide agreements, and to reach the level of 285 within eighteen months (Fossati and Said, 1983). Progress to and especially beyond Coefficient 305 takes many years. Thus the *techniciens* tend to be about the same age as the *ingénieurs débutants* and the *ingénieurs diplomés* in Position II, and, like them, to have most of their careers before them. Yet, their futures are very different because their upward mobility is blocked in the short run and much more limited in the long run. This frustrating situation seems to have become more serious with the rise of *technicien* educational programs that enable technicians to enter the *laboratoires* ladder at a fairly high level but young age.

Credentials and seniority

The structure of opportunity varies not only according to positions and their quality as stepping stones, but also according to the "qualifications"

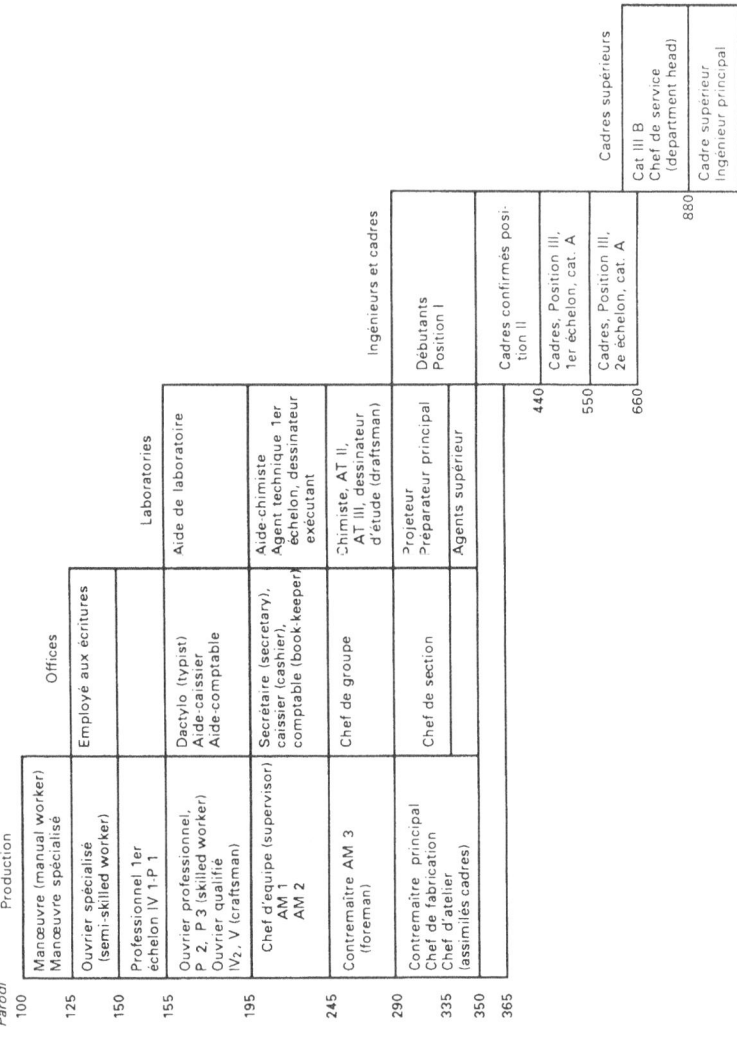

Table 7.1: Occupational hierarchies in French industry (adapted from C. & M. Durand: 1971: 14).

individuals bring to them. Age and sex are important here, but of particular significance in France are educational credentials and seniority. About half of the *ingénieurs* in Positions I and II are *ingénieurs autodidactes*, and most of them will never make the jump to *cadres supérieurs*. This is not simply because of their somewhat higher age, but because French companies prefer to limit middle and higher management positions to *cadres diplomés*. The lack of a diploma also means they would have difficulty entering a different company at the *cadre* level. Thus, *ingénieurs autodidactes* are disadvantaged on both the internal and external labor market, relative to their colleagues with degrees in the same position.

By contrast to *techniciens supérieurs* and *ingénieurs autodidactes*, *ingénieurs diplomés* begin their careers as *cadres* and enjoy substantial opportunities for advancement into the ranks of middle management. Fully 70 percent of them have become *chefs de service* by age 40; by the end of their career, over half have risen to the level of directing an establishment (a factory, for example) or higher (C. and M. Durand, 1971: 45).

Educational credentials determine different career opportunities not only for *ingénieurs diplomés* and *ingénieurs autodidactes*, however, but also for differently schooled groups of *ingénieurs diplomés*. The section on salaries portrays the relationship between diploma and salary. Table 7.2 shows the precise way in which French industrial firms (TELECO in this case) rank schools of higher learning for purposes of hiring, promotion, and pay. Table 7.3 lists several of the largest of those schools in the order of their ranking in Table 7.2, and shows the percentages of graduates working in different areas (hierarchical and functional) in 1975, according to the UIMM (1977) survey of the metal industries. The data in Table 7.3 are uncontrolled for age and so are less indicative than would be figures on final status attainment at retirement, but they give a reasonably accurate picture of the powerful effects of educational credentials.

Comparative studies have frequently commented on the exceptional role of educational credentials in the higher reaches of French organizations. Granick's (1972) study goes farther, highlighting the distinctive character of organizational careers in France by systematically comparing the French, British, American and Russian systems for recruiting and promoting managers. It is worth quoting his conclusions at length:

British, American, and Russian large industrial organizations are all of a fairly "open promotion" type, in which the force of the pre-entrance qualifications is relatively weak. In all three countries, the enterprise management faces the task of considering a very substantial body of men when making promotions to middle and upper managerial posts; the converse of this is that large numbers of junior and middle managers can legitimately aim for promotion to senior positions. In contrast, large French enterprises have a "closed promotion" character and the task of selecting men for promotion according to their performance within the company is greatly

Table 7.2. *French engineering schools by prestige rank*

Category A	
——	Doctorat ès sciences
EP	Ecole Polytechnique
ECP	Ecole Centrale Paris
ENST	Ecole Nationale Supérieure de Télécommunications [Telecom]
ENSAé	Ecole National Supérieure Aéronautique [Sup'Aéro]
ESE	Ecole Supérieure d'Electricité [Sup'Elec]

Category B	
EMP	Ecole Mines de Paris
ESPCI	Ecole Supérieure Physique et Chimie Industrielles
ECL	Ecole Centrale Lyonnaise
ENSERG	Ecole Nationale Supérieure d'Electronique et de Radioélectrique (Grenoble)
ENSAM	Ecole Nationale Supérieure des Arts et Métiers
ISEP	Institut Supérieur d'Electronique de Paris
ENSEM	Ecole Nationale Supérieure d'Electricité et de Mécanique (Nancy)
ENSEEIHT	Ecole Nationale Supérieure d'Electrotechnique, d'Electronique, d'Informatique et d'Hydraulique de Toulouse
IDN	Institut Industriel du Nord de la France (Lille)
CESTI	Centre d'Etudes Supérieures des Techniques Industrielles
ENSMAé	Ecole Nationale Supérieure de Mécanique et d'Aéronautique (Poitiers)
ESIM	Ecole Supérieure d'Ingénieurs de Marseille
ICAM	Institut Catholique d'Arts et Métiers (Lille)
ECAM	Ecole Catholique d'Arts et Métiers (Lyons)
ENREA	Ecole Nationale de Radiotechnique et d'Electrique Appliquées (Clichy)
——	Maîtrise ès Sciences
ENSEEX	Ecole Nationale Supérieure d'Electronique et d'electromécanique (Caen)
ENICA	Ecole Nationale Supérieure d'Ingénieurs de Construction Aéronautique (Toulouse)
ENSM	Ecole Nationale Supérieure de Mécanique (Nantes)
ESEO	Ecole Supérieure d'Electronique de l'Ouest (Angers)
HEI	Hautes Etudes Industrielles (Lille)
ESME	Ecole Spéciale de Mécanique et d'Electricité (ex Sudria)
ISEN	Institut Supérieure d'Electronique du Nord (Lille)
ETP	Ecole Spéciale des TP du batiment et de l'industrie
INSA	Institut Nationale Sciences Appliquées (Lyon, Rennes, Toulouse)
ESIEE	Ecole Supérieure d'Ingénieurs Electrotechnique et Electronique (ex Bréguet)

Table 7.2 (*cont.*)

Category C	
--	Ancienne licence ès sciences
ENAC	Ecole Nationale de l'Aviation Civile
ENSCM	Ecole Nationale Supérieure Chronométrie et Micromécanique (Besançon)
EFRE	Ecole Française de Radioélectricité et d'Electronique
ESERB	Ecole Supérieure d'Electronique et de Radioélectricité (Bordeaux)
EEMI	Ecole d'Électricité et de Mécanique Industrielle (ex Violet)
EEIP	Ecole d'Electricité Industrielle de Paris (ex Charliat)
ISIN	Institut des Sciences de l'Ingénieur (Nancy)
ENI	Ecole Nationale d'Ingénieurs (Belfort, Brest, Metz, St. Etienne, Tarbes)
EPF	Ecole Polytechnique Féminine

simplified. Far more than is the case in the other three countries, the selection job has already been done at the moment of a man's entrance into the firm. (*1972: 197–198*)

Granick points to other distinctive characteristics of French careers as well. Thus, "the truly striking result in comparison with American career patterns is the length of time managers spend in a single post." Moreover, French "managers move between companies only very rarely after the age of about thirty" (pp. 234–235).

It is important to recognize that French credentialism applies largely to entry level positions, especially at the higher levels of the organization (e.g. *cadres*). Subsequently, seniority takes on great significance for career progress and wages. "For industrial wage earners as a whole, the effect of seniority on wages is three times higher in France than Germany . . . Seniority is particularly discriminant for middle management" (Maurice et al., 1984: 266). Maurice et al. explain why:

Though one cannot make broad generalizations for all of French management, these traits are nonetheless the most basic: becoming a manager corresponds to a promotion up the status hierarchy rather than the attainment of greater technical ability. Therefore access to a management position and internal mobility within the category will generally result from the way the company functions and from socialization in accordance with its organizational norms. Attachment to the company tends to become one of the principal criteria for stratification, as evidenced by the greater average length of service; French managers have more seniority compared with their German counterparts. In a sense all the managers, including those who graduated from a *Grande Ecole*, are considered by the company to be "self-taught": success in the company depends on their degree of integration. (*1984: 253*)

For *ingénieurs diplomés* who have diplomas from schools of the same

Table 7.3. *Engineers career positions by diploma status, 1975*

Diploma by TELECO ranking	General (top) management	Administration and sales	Research and design	Production, methods, tests	Other	Total	N
Polytechnique	38	32	19	6	6	101	1,924
Centrales	18	31	26	19	6	100	4,502
Telecom, Sup'Aéro, etc.	12	24	47	13	3	99	1,764
Arts & Métiers	12	17	23	41	6	99	9,686
Other *polyvalent*	7	36	23	28	6	100	8,700
Specialized: mechanics	4	18	31	42	6	101	3,156
Specialized: electricity	4	38	24	29	5	100	4,776
Specialized: electronics	3	22	43	27	5	100	4,474
Facultés des Sciences	3	34	39	18	6	100	6,868
Part-time (CNAM etc.)	4	27	35	29	5	100	5,545
BT, BTS, DUT, etc.	3	39	22	31	6	101	16,393
Autodidactes	3	57	13	23	5	101	44,773

Source and notes: UIMM, Ingénieurs et Cadres des Industries Métaux, 1977, pp. 40–61. Centrales is plural because figures also cover less prestigious but small sister, Central Lyonnaise. The etc. after Telecom, Sup'Aéro stands for Ponts-et-Chaussées and Génie Maritime; these four elite schools share a common entrance exam. I have left out the Ecoles des Mines, and the combination ESE–ENSI–ICAM, because in both cases the figures average graduates from schools of quite different ranks (different groups); these five schools account for another 14,416 *ingénieurs diplomés*. I have also left out those with diplomas from numerous smaller schools. They would add 9,715. The specialized schools combine those of varying rank, so their relative positions above are somewhat arbitrary. The N is for the number of graduates in the metal industries only.

prestige level, seniority becomes a major determinant of their career and salary positions.

This study has mentioned several times the role of the educational system and its production of the holders of various diplomas. In particular, chapter 5 emphasized the expansion of the *petites écoles d'ingénieur*. It is important to recognize, however, that the effects of such shifts are mediated by the autonomous recruitment policies of employers. In a careful statistical analysis of recent changes in the educational qualifications of German and French workers and the employment of technicians, engineers and higher administrators, Paul Windolf (1983) concludes that the difference in the percentages of highly qualified technicians, engineers, and managers – traditionally higher in France – increased during the 1970s. This occurred despite Germany's educational system catching up with France's in the production of holders of higher degrees (by industry). In his words, these "national differences are better explained by the different strategies of recruitment than by differences in the occupational structure" (1983: 127).[5]

Lateral mobility, the division of labor, and job security

It is hardly surprising that *ingénieurs autodidactes* do not change companies often. A major criterion for their promotion to the position of *ingénieur* is faithful service for many years, and their lack of a higher diploma puts them in a relatively weak position on the external labor market. This is probably one reason that a higher percentage of them than of *ingénieurs diplomés* express an interest in one day establishing their own business.[6] But *ingénieurs diplomés* are also quite stable. According to the 1977 FASFID survey, 45 percent of them, excluding *ingénieurs débutants*, have always worked for their current employer. An additional 27 percent have worked for only one other employer. By contrast, French manual workers change employers frequently. According to Zeldin (1979: xv) 42 percent of them "had changed their trade four or more times" in the 1950s, and 24 percent had changed their place of work eleven times.

Granick (1972: 294ff) points out that both the "intercompany mobility rates" of managers and the intracompany "turnover rates" among managerial posts are much lower in France than in Britain or the United States.[7] The reason, he suggests, is the closed promotion system in France. Since the ranking of diplomas is virtually the same throughout French industry, the individual engineer has little to gain by shifting to another firm. By contrast, in societies where many are eligible for promotion, those not chosen at one firm may want to try their luck elsewhere.

One factor in intercompany mobility is job security. Traditionally, French firms, as part of their paternalism, have been reluctant to "lay off" workers and very averse to discharging *cadres*. But as Granick (1972: 378) notes, "the combination of a change in managerial values toward dismissals and of the merger movement in large enterprises had resulted in an entirely new phenomenon for post-war France: namely the beginning of the unemployment of *cadres*." Between 1965 and 1973 the number of *ingénieurs*, *cadres*, and *techniciens* seeking work grew twice as fast as the number of job openings, and *cadre* unemployment grew accordingly (Groux, 1983: 39). Some analysts (Dulong, 1971: 172) of the events of May–June, 1968 – the nationwide general strike by students and workers – attribute the participation of some *cadres* to the emergence of unemployment among *cadres* and the accompanying disillusionment concerning management's commitment to them. Perhaps in response to such agitation, "the law of 13 July 1973 states that dismissal must be for 'real and sufficient cause,' and any firm that fails to meet this test may be required to pay indemnities to the dismissed employee" (Boltanski, 1987: 377).

Nevertheless, the unemployment of *cadres* has grown, from 14,000 in 1971 to 40,000 in 1975 to 64,000 in 1981 (Groux, 1983: 39). This was a period of increasing unemployment among all French workers, so the proportion of *cadres* unemployed remained considerably below the general average. Yet, the psychological blow to the meaning of a *cadre* position seems to have been substantial and enduring.[8] If the fear of being black-listed as an untrustworthy employee who abandons his employer has receded with the decline of paternalism, the fear of not finding another job has increased. At the time of the interviews the unemployment rate for *ingénieurs* and *techniciens* alike was 2.4 percent (vs. 5.9 for semi-skilled workers) and rising. Mobility was further discouraged by the growing number of employed spouses. Whatever the reasons, in 1977 84 percent of employed *ingénieurs diplomés* reported intending to stay with their current employer.

To be sure, job security and turnover rates vary by industry. As elsewhere, the textile and metal-working industries in France face stiff competition from abroad, and are struggling to survive. PAMPCO had lost money the year before the interviews, and had instituted a freeze on *cadre* hiring. The electronics industry by contrast was prospering, and TELECO's Simulator Division was aggressively recruiting young engineers. Company differences were evident also in the turnover rates; only 23 percent of the engineers interviewed at PAMPCO had ever worked for any other company, as opposed to 43 percent at TELECO, although the greater opportunities in the Paris region should not be neglected. Such company differences will come up again in the interpretation of these engineers' industrial and political attitudes.

At the same time, a note of caution is in order. In distinguishing among industries, it is difficult but important to disentangle the effects of technology from those of newness. High technology industries are new industries, and new industries, whatever their technology, tend to face different kinds of problems and opportunities than old industries (Stinchcombe, 1965). Because new industries offer unusual opportunities for capturing emerging markets but unusual risks stemming from intense competition and uncertainties about demands and costs, they tend to be unstable. The market within the industry at large may be strong, but the future of an individual firm may be shaky.

A second note of caution concerns the meaning of unemployment. While a structural trend towards greater employment instability may be raising the unemployment rates of French engineers, and a temporary business recession exacerbated matters in the mid-1970s, the general labor market for engineers appears very strong and likely to remain so. Indeed, there are several predictions of increasing and serious shortages of engineers and technicians in France, especially within the electronics sector (*Nouvelle Usine*, 1982). At the same time, the very existence of such predictions makes the experience of even temporary unemployment all the more disconcerting to the individual engineer.

Salaries

French *ingénieurs* and *cadres* complained frequently in the mid-1970s about an alleged relative decline in their salaries, and these complaints were echoed loudly in the business press. Spokesmen for both the major *cadre* union, the CGC, and the Communist union for *cadres*, the UGICT (CGT), decried the narrowing of *cadre*–worker salary differentials (Vincent, 1977). Marxist intellectuals were relieved to have what seemed to be evidence of the long anticipated proletarianization of the new middle class. It turns out, however, that the existence of such a decline depends on the base year chosen for comparison. It tended to be a decade, but as Table 7.4 shows, going back two decades or more yields a different picture.

It is hardly surprising that such figures are subject to dispute, for they reflect only pre-tax salary, not net income after taxes and family allocations, etc. Moreover, *ingénieurs* have lost some ground to *cadres supérieurs administratifs* within the category of *cadres supérieurs*, slipping from a coefficient of 0.93 in 1956 to 0.87 in 1975. On the other hand, the growth of

Table 7.4. *Evolution of salary ratios*

	1950	1951	1954	1963	1967	1969	1971	1973	1976	1977	1981
Ratio											
cadres supérieurs:ouvriers	3.4	3.8	4.1	4.2	4.5	4.2	4.3	4.0	3.8	3.7	3.5

Sources: Centre d'étude des revenus et des coûts (November 1977: 57). (The 1981 figure comes from Groux, 1983: 48.)

the salary differential between *cadres supérieurs* and *ouvriers* until 1968, followed by its return to its 1950 level in 1980, coincides with both a major expansion and then contraction of the economy. It is quite common for salary differentials in capitalist democracies to increase in times of prosperity and decrease in times of depression. In other words, the decline may well have reflected less a secular decline with the evolution of capitalism than the down phase of an ebb and rise linked to the business cycle.

Even more important is a point about the age factor. More than most other occupations, French engineering careers involve large increments in salary. This means that if the age distribution shifts downward, average salaries decline, even though average lifetime earnings may not. We saw in the previous section that just such a shift in the age structure of the French engineering population seems to have occurred in recent years. Finally, there is some evidence that senior engineers have maintained their relative position within the national salary hierarchy. Boltanski (1982: 225) reports that an *ingénieur en chef* (the next rank above a *chef de service*) earned eight times the salary of a manual worker in 1951. I calculate that in 1976 an *ingénieur en chef* earned 7.6 times what a manual worker earned.

It is beyond the scope of this analysis to determine whether the recent relative decline in engineers' salaries represents a temporary adjustment to changes in the age structure, a passing response to cyclical supply or demand factors, a general deskilling of engineering (see chapter 5), or a general broadening of the base of the engineering hierarchy. However, if it is the result of cyclical factors or age distribution shifts, it is certainly premature to infer any "proletarianization." And *even if* the proportion of higher ranking to lower ranking engineers has permanently declined, it is misleading to speak of proletarianization as long as the ratio of higher engineers to all industrial workers does not decline. In fact, it has risen.

One reason the graduates of the most prestigious *grandes écoles* have not been hurt by any decline in the average value of an engineering degree accompanying the recent credentials inflation is that the elite schools have successfully resisted pressures to expand. Table 4.1 showed that while the number of students in *petites écoles* increased by more than 200 percent between 1958 and 1978, the number of students in the elite *grandes écoles* increased by less than 50 percent.[9] Some may argue that this pattern of change in the production of engineers reflects developments on the demand side, but the work of Collins (1979: ch. 1) on credentialism suggests otherwise. So do reports on the politics of French higher education and the effectiveness of the alumni associations of the elite *grandes écoles* at preventing expansion of their schools (see chapter 8).

Even with the recent expansions in engineering education, France still produces fewer graduate engineers per year than other comparable countries. In 1980, the figures were as follows: France, 11,000; England,

12,500 (chartered); Germany, 25,000 ("diploma" and "graduate"); and the U.S.A., 87,000 (bachelors, masters, and doctors). French salary differentials have long stood out as greater than those of other industrialized nations. According to a Common Market survey reported in *L'Expansion* (March, 1978: 82) the ratio of the average salary of a male worker (unskilled or semi-skilled) to that of a *cadre supérieur* varied from a low of 1:3.6 in Belgium and Germany to a high of 1:5.4 in France.[10] Fores (n.d.) claims that in 1970, the average French engineer earned about five times as much as the average French industrial worker, a salary differential of twice that between British engineers and British industrial workers.

The salaries of French engineers rise considerably over the course of their careers. FASFID's 1980 issue of *ID* reported that the average *ingénieur diplomé* aged 60–65 earned three times what an *ingénieur débutant* earned. Educational credentials strongly affect salary growth curves, although again, the different growth rates of different groups of schools complicate the interpretation of the data. Figure 7.1 presents a chart that appeared in the French business journal, *L'Expansion* in June, 1977. From this chart, it is clear that the average salary of the graduates of elite (A) engineering schools is higher than the comparable salary of their *petite école* counterparts (C) at the beginning of their careers, and rises at a faster rate to a point where it is almost 50 percent higher at age 52. These differences are partly due to the fact that the *grande école ingénieurs* progressively occupy higher positions than other engineers. But only partly, for the starting salaries of *ingénieurs débutants* (Position I) vary consistently by education, even within the same firm.

Table 7.5 shows the official guidelines for PAMPCO and TELECO. It is evident that these companies pay in terms of diploma status. At TELECO in particular, diploma status determines a starting engineer's salary. The best paid in Groups III and II earn less than the worst paid in the next higher group. At PAMPCO, by contrast, the best paid in Group III is paid more than the worst paid in Group I. Thus, at the more paternalistic PAMPCO, management retains more flexibility to adjust rewards according to other personal qualities and performance. Also noteworthy is the considerable range of starting salaries from top to bottom. The best-paid *ingénieurs diplomés* at PAMPCO earn up to 57 percent more than the worst-paid, as opposed to 27 percent more at TELECO. Obviously the market situation of even so apparently homogeneous a group as *ingénieurs diplomés débutants* in the same company varies widely.

It is, however, the salary differences between the two companies that are of most interest.[11] Table 7.6 displays the average salaries of certain categories of *ingénieurs* at PAMPCO and TELECO. It is evident that the PAMPCO *ingénieurs* do better than their counterparts at TELECO. This is surprising, because the salaries of *cadres*, aged 30–34 in the Paris region,

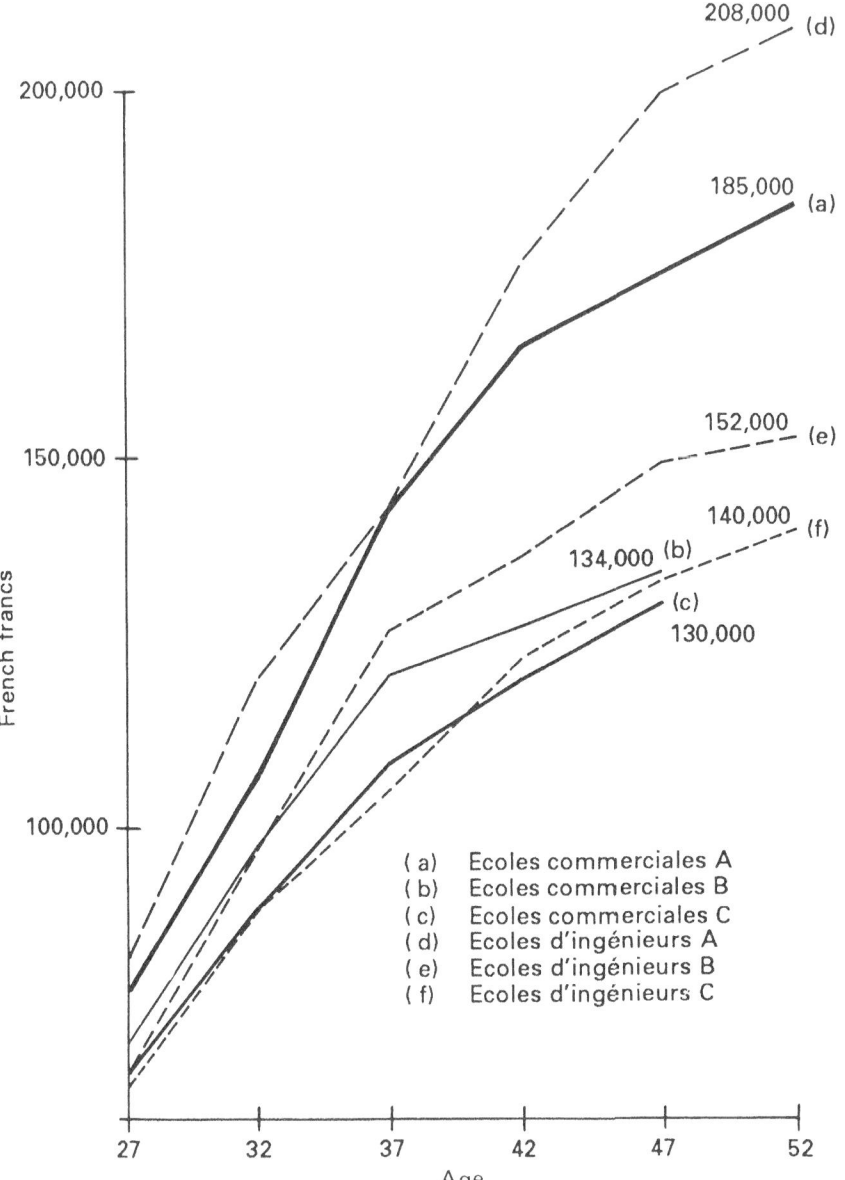

Figure 7.1: Salary profiles by school rankings (Source: *L'Expansion*, June, 1977)

Table 7.5. *Starting salary ranges by diploma status*

PAMPCO: Annual salaries, 1978		
Diploma groups	Lowest	Highest
Ia (highest ranked schools)	65,000 FF	80,000 FF
Ib	62,500	77,000
IIa	58,000	71,500
IIb	56,000	69,000
IIIa	52,500	65,000
IIIb	51,000	63,000
IV (*nondiplomé*)	46,000	56,500
TELECO: Annual salaries, late 1977		
Diploma groups	Lowest	Highest
I	69,000	73,000
II	63,500	68,200
III	58,000	62,800
IV (DUT or BTS)	46,300	57,100

Source: Company documents.

Table 7.6. *Average salaries at PAMPCO and TELECO, 1978*

Age	PAMPCO	TELECO
	Ingénieur II	*Ingénieur* II
30–34	98,406 FF	95,472
	Ingénieur IIIA	*Ingénieur* IIIA
35–39	132,000	124,800
	Ingénieur IIIB	*Ingénieur* IIIB
45–49	184,000	166,000

where the cost of living is high, average about 5,700 FF higher than those of otherwise similar *cadres* in the Lorraine. Of course, the engineer at TELECO does not compare himself to engineers at PAMPCO, but rather to norms for the nation or his industry. This is easy to do in France. Every May, *L'Expansion*, a magazine modelled after the American financial monthly, *Fortune*, publishes a special 30-page review of *cadres'* salaries. Included in this review is a formula, under the heading "calculate your salary," which reveals the averages – hence what one "should be" earning – for a wide combination of jobs, ages, educations, regions, etc. Almost all TELECO *ingénieurs* were earning less than their hypothetical counterparts – 10 to

15 percent less in many cases. Data from the last relevant FASFID survey (1977) tend to confirm this portrait.

More importantly, the salaries in electronics tend to run about 4,000 FF below the average for French industry for this age group, 5,000 FF below it for "ordinary *ingénieurs*" (all below *chefs de service*), although this is partly "because" so many work in design. Because of the overlaps in these categories, precise comparisons with TELECO are impossible. Nevertheless, it is striking that while the 1977 average for electronics *ingénieurs diplomés* aged 30–34 was 93,000 FF, the comparable figure at TELECO was 83,000. While the industry-wide average for ordinary *ingénieurs diplomés* was 85,000, the almost comparable figure at TELECO (*Ingénieur* I and II) at the Simulator Division was 77,000 FF (84,000 for those aged 30–34).

In examining labor markets, it is important not to ignore the people that enter them, and the processes by which they reach them. Although it is beyond the scope of this study to analyze all aspects of self-selection and selection by others for various categories of French engineers, the following section does address the issue of recruitment and socialization.

Recruitment and socialization

Theories of professionalization suggest that the shared experiences of professional schooling foster a "professional" orientation to work and even occupational commitments and solidarity. In the process, they assume that the vast majority of the profession's members enter it by way of a similar professional school, that the principal function of schooling is to provide professional training, and that the function of the subsequent hiring process is to identify the most professionally qualified candidates. In the case of French engineers, such assumptions warrant careful examination.

By defining French engineers as *ingénieurs* and *techniciens supérieurs*, this study may seem unfairly biased in favor of an argument showing that many French engineers are not graduates of engineering schools. Therefore, the following analysis excludes *techniciens supérieurs*, dealing only with practicing *ingénieurs*. Data are available only for the "metal industries" in the private sector. These industries employ only about two-fifths of all French *ingénieurs*, but cover a wide range of core traditional and science-based industries, from forges and foundries to automobiles and electronics.

Recruitment to the occupation

Only about three-fifths of the practicing *ingénieurs* in France's metal industries are graduates of engineering schools. The remainder became *ingénieurs* by way of a variety of other routes. Table 7.7 shows the percentages of *ingénieurs* from different educational backgrounds and the trends in these percentages since 1956.

Table 7.7. *Metal industry ingénieurs by educational background and year*

Educational background	1956	1962	1970	1975
1 Graduates of full-time engineering schools	67%	60%	58%	59%
2 Graduates of other professional schools	1	2	2	2
3 Graduates of part-time engineering programs	—	—	4	5
4 Graduates of schools for technicians	9	9	7	11
5 Unschooled *ingénieurs*	23	29	30	23
	100%	100%	101%	100%

Source: Ingénieurs et cadres, 1977, a report by UIMM. The UIMM does not distinguish *ingénieurs* from *cadres*, but does classify *cadres* according to their principal type of work: sales, administration, research, etc. The above figures represent my calculations for all those working in the four obviously technical types of work.

Two characteristics of this table bear emphasis. The more striking is the stable proportion of *ingénieurs* coming into the occupation via the different routes. More specifically, *ingénieurs diplomés* (line 1) have not gained ground, despite the great expansion in their production during the 1950s and 1960s. The other noteworthy characteristic is the variety of routes by which *ingénieurs* without diplomas have become *ingénieurs*. The graduates of "other" professional schools are graduates of business, law, and similarly prestigious university-level programs. Like *ingénieurs diplomés*, they are hired as *cadres* on the strength of their diplomas. Close to the *ingénieurs diplomés* in age and status, most of them work in the least technical of the technical functions, sales engineering. By contrast, the graduates of the part-time engineering programs are *ingénieurs diplomés*, but are somewhat older and have not shared the potentially formative experience of full-time professional schooling. More importantly, most have worked many years as non-*cadres*, becoming *cadres* only after obtaining their diplomas.

The "unschooled" and graduates of technical schools – 34 percent of the total in 1975 – are *cadres autodidactes*, former technicians and workers who lack higher educational degrees and have thus acquired *cadre* status by internal promotion rather than educational credentials. Even though French sources often define all *ingénieurs nondiplomés* as *autodidactes*, there are really two quite different career paths here. The "graduates of the schools for technicians" are the holders of high school or junior college technical degree, such as *Brevets Techniques*, *Brevets de Techniciens Supérieurs*, and *Diplômes d'Université Technique*. They tend to be assimilated into the ranks of *cadres* at an earlier point in their careers than the *ingénieurs autodidactes* who have less formal schooling, and thus, by both educational level and by age, often have more in common with the *ingénieurs diplomés* than they do with many *ingénieurs autodidactes*. The latter, however, are far from an educationally homogeneous group; some have the *baccalauréat* (*Bac*), others only a primary school education. Many have lower vocational level training certificates and have acquired their higher technical skills by working for years as skilled workers and foremen. Surprisingly, in an age of emphasis upon technical training and educational credentials, these unschooled *autodidactes* represent almost one-quarter of *ingénieurs*, a figure somewhat lower than that for 1962 and 1970 but no lower than that for 1956.

Whatever their formal schooling, the *ingénieurs autodidactes* represent a sub-stratum within the ranks of *cadres*. Part of the reason for this is their different position in the internal and external labour market. Another reason is that among lower ranking *ingénieurs* they represent an older group of former workers whose attainment of *cadre* status by promotion represents something of unusual significance in France. As Maurice et al. (1977: 770) explain, "In France, the fundamental change of status in a company is the *passage-cadre*, a passage which . . . represents . . . a shift in allegiance away from the union and towards the company." Maurice et al. show that in Germany, by contrast, there is a much finer gradation of hierarchical levels so that a passage between loyalty to union and loyalty to company is not sharply experienced during a career; also, promotions are based more on training and competence than demonstrated loyalty and "potential." In France, *ingénieurs autodidactes*, company engineers lacking engineering degrees, are seen as a particularly grateful, hard-working, and loyal group. Not surprisingly, the *ingénieurs diplomés* often view them as technically under-qualified but over-qualified in terms of the kinds of character traits that management favors: loyalty, hard work, etc. In the words of a personnel officer at TELECO: "There is an aristocracy of *cadres diplomés*. *Autodidactes* – the word itself is pejorative – and those promoted from the ranks are often called 'mulattos'".

In short, a large and persisting proportion of French engineers – 39 percent – do not share the major occupational socialization experience to which the professionalization theorists refer, professional schooling. Nor do they share a single alternative route, such as apprenticeship. The significance of the fact that French engineers are recruited from a variety of educational backgrounds will become more evident when we examine the unionization and politics of these workers in subsequent chapters.

What about the occupational socialization of the *ingénieurs diplomés*, all of whom have attended an engineering school? To be sure, the rigorous educational experience is shared. The identification, however, is with *grande école* graduates in general, whether technical or not, and with fellow graduates of the same school, whatever line of work they pursue later. An important factor in the broader identification is that French engineering schools, themselves partly sponsored and financed by private industry, actively promote a managerial rather than professional image of an *ingénieur diplomé*. Thus, the Director of one *grande école* in his introduction to the semi-official *Les Ecoles d'ingénieurs en France*, contrasts the "general competence," "hierarchical responsibilities" and multi-occupational managerial careers of French engineers to the specialized character of American engineers and the technical staff character of English engineers. In his words:

> During his career an *ingénieur* can change occupation (*métier*). A frequent example: he begins in a *métier* which is devoted essentially to research or production, then orients himself more and more towards management . . . Very few specialize and attempt a career of pure engineering on the model of Gustav Eiffel. (*Laffitte, 1973: 19*)

In documents like this, in speeches, in classes, and in numerous more subtle ways, the students in French engineering schools are discouraged from developing a strong attachment to a particular location in the division of labor or those with whom they share that location.[12] In short, they are encouraged to define the role of an engineer as that of an adaptable organization man, not a professional.

Recruitment to the firm

The young *ingénieur diplomé*'s search for a full-time job usually begins during the last few monjths of his national service. Nearly all of them – 86 percent, according to a 1983 survey by the Association pour l'emploi des cadres (APEC) – send unsolicited inquiries to any company that interests them (*Le Monde de l'Education*, 1984: 46). (The exceptions are the students of the most prestigious *grandes écoles*; companies come looking for them.) Almost as many of these engineers reply to company recruiting advertisements in newspapers, trade journals, and alumni association bulletins. Nevertheless, according to a 1979 survey, 60 percent of *ingénieurs débutants* point to personal or professional contacts, the latter often established during one of their summer on-the-job training practicums (*stages*) in industry while at engineering school, as being important in their final choice of an employer (*Le Monde*, 1984).

These searches for a first job resulted in about five employment offers for the average candidate in 1983, and more than ten for the graduates of better schools. Surveys concerning the motives for choosing among these offers suggest that starting salaries were important, but less so than the opportunities offered for further training and career development. The reputation of the firm is also important, especially among engineers seeking jobs in the high technology sector (*Le Monde*, 1984: 46). The reasoning here seems to be that many young engineers, especially in the rapidly growing but unstable high technology sector, view their first job as an experience-gaining trial run in the world of work. "And the experience acquired over two or three years strengthens one's position on the labor market" (1984: 35).

It is not only the schools that encourage their students to define the role of an engineer as that of an organization man rather than of a professional. The companies do too, and this is evident from their procedures for hiring *ingénieurs diplomés*. In a valuable study, Benguigui (1981) makes the following points about the hiring process for *cadres*. First, French companies are extremely careful about whom they hire as *cadres*; hiring is usually centralized in a special *cadre* personnel service at company headquarters, and often involves six to eight weeks of investigation. Second, the various tests and interviews to which the candidates are subjected are designed to ascertain psychological traits (especially adaptability) and social abilities. These tests range from personality inventories to handwriting analyses, but are rarely technical; the interviews are conducted by personnel managers and *cadre* employment service agents who normally lack the training to judge a candidate's technical qualifications. Benguigui quotes one executive as saying: "We recruit *ingénieurs* not for a specific job but for a career (*une vie professionnelle*); we don't seek acquired knowledge, but a personality." Finally, the competitive, rigorous, and yet non-technical character of the hiring process, typically including national searches through newspaper advertisements that specify an unlikely array of skills and abilities, tends to make the successful candidate feel lucky, appreciative, loyal, and under continuing pressure to prove himself worthy. In brief, the hiring process identifies and selects *ingénieurs* whose values are compatible with those of a particular company's management, and at the same time reinforces the employer's power to define the proper role of an *ingénieur* in the ensuing relationship between them.

The process Benguigui describes closely resembles that followed at PAMPCO. There, one *cadre* presonnel manager told me that PAMPCO recruits only *ingénieurs débutants*, and "hires a man, not a diploma." He emphasized that "the best school is the company, especially the production side of it." The personnel office chooses a subset of candidates on the basis of records, psychological tests, and long background investigations, interviews six, and sends "the two best" to unit directors for final selection.

At TELECO, however, the process and premises were quite different. The Simulator Division had its own personnel manager, a sociologist, and was aggressively recruiting engineers. Less concerned with making the best possible choice, TELECO gave much less time to background investigations. The reason is that the company viewed the first two years on the job as a probationary period, and expected to make a further selection after that. This strategy proved quite compatible with the attitude of many young engineers in electronics that their first job should be at a big, well-respected company, after which they would be in a stronger position to seek the kind of long-term job they wanted, perhaps in the provinces.

Labor market attitudes at PAMPCO and TELECO

The previous sections have examined several aspects of the labor market for PAMPCO and TELECO engineers. But how did these engineers feel about their careers to date? What were their ambitions, and how did they view their chances of realizing them? Did they feel secure in their present position? Would it be difficult to find another job if necessary? How did they feel about their salary and about the company's salary policy in general? This section examines the labor market situation of the PAMPCO and TELECO engineers as they perceived and evaluated it.

Careers

It seems clear that, from the schooling and recruitment of French engineers to the structuring of their career opportunities, French engineers are encouraged to define success as promotions out of purely technical work and into at least middle levels of management. It remains to be seen, however, whether or not many engineers, especially at TELECO, accept management's definition of success. Professionalization theory suggests an occupational culture with its own criteria of success – criteria that emphasize peer recognition for technical expertise and achievement. Proletarianization theory implies a greater concern with pay and security than with recognition by peers and management.

Each engineer was asked, 'When you consider the way your career has gone up to now, are you pretty satisfied, or do you have some regrets or reservations?" Sixty percent mentioned their satisfactions or regrets regarding the work they had done, but 80 percent discussed their promotion experience. Moreover, while most respondents expressed satisfaction with the content of the jobs they had done, only 17 percent said they were pleased with their career progress.[13] What is meant by satisfaction or disappointment in this context? The following comments are revealing.

When I started here I anticipated more rapid advancement, so I am disappointed. [a PAMPCO *technicien*]

I've gone about as far as is reasonable for my age. [a PAMPCO *ingénieur autodidacte*]

I am pleased to have already changed my position. That means I am advancing. [a PAMPCO *ingénieur diplomé*]

I felt dissatisfied with the pace at which my career was evolving until I was finally named *chef du groupe* last year. It's a professional satisfaction to be recognized and given responsibility. [a PAMPCO *ingénieur diplomé*]

I still don't regret transferring to production, because it has allowed me to obtain a promotion more rapidly. The work is dirtier and more constraining, but it has enabled me to advance my career faster than if I had stayed in design. [a PAMPCO *ingénieur diplomé*]

This last comment is noteworthy in that it shows a favorable attitude towards a job that would be regarded as dirty and unsatisfying were it not a good stepping stone to more desirable jobs, desirable here meaning essentially higher in the bureaucratic hierarchy. The following comments are by TELECO engineers.

I don't think it's possible to attain this level any faster than I have. That's satisfying. [a TELECO *chef de service*]

I could have risen farther. I regret getting stuck in a corner under an *ingénieur autodidacte* at a previous research job. I lost two or three years. [a TELECO *ingénieur diplomé*]

I had no prior expectations except to keep moving up and into new things. But I'm satisfied to have advanced as fast as I have. [a TELECO *chef de service*]

I regret not having gone as far as the others who entered when I did. [a TELECO *ingénieur diplomé*]

From these responses, and many more like them, it is evident that these engineers view success as upward movement in the bureaucratic hierarchy. However, upward movement alone is not enough. Since the very structure of French engineering implies upward mobility over time, what distinguishes the more from the less successful is the speed at which they advance, relative to their cohort. For these engineers success is not recognition by their peers as superior engineers, nor is it simply higher salaries. In a question about their ambitions, only 4 percent of the engineers emphasized salary and security concerns, while 43 percent emphasized their careers.[14] Rather success is relatively rapid movement out of technical engineering and up the bureaucratic hierarchy.

Further evidence for this interpretation of the meaning of success to these French engineers emerges from an examination of the relationship between the actual pace of career advancement and the evaluation of a career as satisfying or disappointing. For the *ingénieurs diplomés* (here including *chefs de service*) at each company, regression analysis was used to plot salaries (nine levels) by years employed as an engineer. The resulting curves reveal very similar patterns of steady (although slowing) progression upward. Each engineer was then classified as "slow," "average," or "fast." Table 7.8 shows

Table 7.8. *Career satisfaction by relative career performance*

Career satisfaction, PAMPCO				
Relative career performance	Disappointed	Routine	Pleased	Totals
Slow	75%	25%	0%	30.8% (8)
Average	33	67	0	46.2 (12)
Fast	50	17	33	23.1 (6)
Totals	50 (13)	42.3 (11)	7.7 (2)	100 (26)
Career satisfaction, TELECO				
	Disappointed	Routine	Pleased	Totals
Slow	36%	27%	36%	32.4% (11)
Average	36	36	27	32.4 (11)
Fast	33	42	25	35.3 (12)
Totals	35.3	35	29.4	100 (34)

the relationship between the actual rate of career advancement and the engineer's evaluations of his career.

This table reveals two significant company differences. First, while there is a fairly strong relationship between "relative career performance" and "career satisfaction" level at PAMPCO, there is none at all at TELECO. The reason, I suspect, for the weaker relationship between career performance and career evaluation at TELECO is that relative performances there are more predetermined by credentials and seniority. The very notions of fast and slow are less salient. The above classification of relative career performances no doubt suffers from the failure of the regression analysis to control for the prestige level of the diploma. Unfortunately, there are simply too few cases to do so.

Other data, however, support an interpretation that sees the TELECO engineers as less preoccupied with their careers. When answering the career satisfaction question, only 25 percent of them focused exclusively on their promotion record, as opposed to 51 percent at PAMPCO. Only 21 percent of the TELECO engineers described their career goals in terms that I would characterize as highly ambitious, compared to 36 percent at PAMPCO.

Second, the table reveals that at PAMPCO only 8 percent of the engineers describe themselves as satisfied with their careers, while 50 percent say they are disappointed. At TELECO, by contrast, almost as many (29 percent) report being satisfied as do being disappointed (35 percent). Fifty percent of the PAMPCO engineers also report sometimes wishing they had chosen a different occupation, as compared to 33 pecent at TELECO. The TELECO engineers are also more optimistic about the future. Sixty percent of them rate their chances of achieving their career goals as good, while only 31 percent at PAMPCO agree. Only two engineers at TELECO appeared downright pessimistic about their career futures, as opposed to fully a third of the PAMPCO interviewees. No doubt this is why more PAMPCO engineers – 50 as opposed to 33 percent – wished they had pursued a different occupation in life.

It would be a mistake, however, to interpret the slightly lower hopes for and concern about movement up into management at TELECO as a turn towards professional criteria of success. A good many unambitious TELECO engineers express an interest in lateral movements into such non-technical functions as sales, partly out of concern about becoming technically obsolete eventually. Rather, what is evident at TELECO is a more casual outlook on careers of any type. As one TELECO *ingénieur* said in reply to the question about where he hoped to be in five to ten years: "To be frank, I don't care much. Personally, I would look for something that allowed me to live, but I am not looking *a priori*, either for higher positions or honors . . . I think that work is not a fundamental element of my life. That's all." Another: "I consider my work as a means of earning my living, although I would like it to be as interesting as possible." To be sure, such comments were rare, even at TELECO, but they are indicative of a slightly more detached attitude towards both career and employer. In the context of a company that stressed educational credentials and of a strong external labor market, there is a certain logic to a more detached and casual attitude on the part of the TELECO engineers, even where less selective recruitment to the company has not produced it. Such attitudes may help explain some of the differences in collective action and politics that appear in later chapters.

In speaking of "engineers" in general, it is important not to overlook the variations among the different categories of them. When asked their chances of achieving their career ambitions, only 15 percent of *techniciens* say they are good, in contrast to a third of *ingénieurs autodidactes* and over one-half of *ingénieurs diplomés* (including *chefs*) and *cadres administratifs*. Only 11 percent of *ingénieurs diplomés* – and none of the *cadres adminis-tratifs* – rated their chances as poor, in spite of the much more widespread disappointment among them with careers to date. This faith that they will be rewarded for their service distinguishes these "trusted" *cadres* from their associates on the other side of the *cadre* barrier. Subjectively perhaps even more than objectively, the *techniciens* occupy a different place in the internal labor market, a place they define as one of "blocked mobility."

Security and labor market

Each interviewee was asked how secure he felt in his present job, and why. Surprisingly, the distribution of responses were very similar at the two companies, even though the old industry firm, PAMPCO was facing a much more difficult market for its products than TELECO was. In fact, 15.5 percent of the engineers at TELECO, as compared to 13.5 percent at PAMPCO, expressed some insecurity about their futures with their present employer. About half of the respondents at each firm reported feeling quite secure in their jobs, the remainder falling somewhere in between. However, if the distributions at the two firms were similar, the reasons given varied considerably. The most common explanation at PAMPCO was, in the words of one respondent, that "PAMPCO doesn't lay off people, even in bad times – *cadres* and *ingénieurs*, that is – because it is large and pretty stable, and because of its type of management." In short, the "trusted workers" at this paternalistic firm had confidence that their employer would not abandon them. Although several respondents at TELECO expressed a similar sentiment, the most common explanation there for feeling secure was that the local unit was doing well. The second most common reason for confidence at both companies was much more individualistic. These respondents pointed to their own seniority, needed skills, or good reputation as a worker, as well as the size and strength of the company. As one PAMPCO *ingénieur diplomé* put it: "this is a good, big company. If you do a good job, I believe one is still secure." Another said, "I'm secure because of my specialty."

Most interesting, however, are the inter-firm differences among the engineers who qualified their affirmative answers or admitted to feelings of insecurity. The following comments are illustrative of the responses at PAMPCO.

I think that my job is becoming less and less secure, for two reasons. First, there's the national recession and the bad situation of our company. They are going to have to lay off some people. Second, management today is less committed to the security of *cadres* than it was in the past. It is less paternalistic – in the good sense of the term – than it was. The new technocrats are getting the company into difficulty and yet will fire us *cadres*, something management would not have done in the past.

I don't feel secure. They can lay off *cadres* like anyone else. One must not deceive oneself. They will cut whole divisions if they need to. I feel particularly vulnerable in this department because its activities are somewhat peripheral to the company's main activities.

Such concerns are hardly surprising, given the state of the market for PAMPCO's products and the growing competition from "third world" producers. Still, they do reveal some resentment at the loss of former privileges. More surprising, however, are the insecurities expressed by some of the TELECO engineers.

There could be layoffs in four or five years, even of *ingénieurs*. The company is weak in exports. There's a lot of talk but not much action and even fewer results. Even the temporal electronic exchange will be too old to sell abroad in a few years, and the company is not developing a successor rapidly enough.

Security here depends on how the company does, on decisions by our main client, the P.T.T. It also depends on good management. I'm a bit pessimistic, myself. It's very possible I'll be laid off one day.

I feel quite secure for at least the next two or three years, but beyond that it's much less clear. Five years from now, the number of people working in production will be half what it is today. Not because machines are replacing people – which would be good for maintenance types like me – but because the new versions of our product require fewer components to do what they do. It simply takes fewer hours' work to produce the telephone exchange capacity desired by a client.

Although such remarks were far more common in the factory for telephone exchanges than at the Simulator Division, they are striking testimony to a kind of instability that is often found in high technology industries, despite a general expansion. Electronics and telecommunications may be industries of the future, but several factors introduce considerable uncertainty into the world of careers within them. As new industries, and as knowledge-intensive ones with low start-up costs, there is or suddenly can be severe competition among firms within them. Many firms are struggling to be among the few that remain at the end of the inevitable "settling out" process. For high technology industries, there are the additional complications arising from rapid changes of technology and from miniaturization. Moreover, in the professional products sector, there are often only a few potential customers for a highly customized product, so that a great deal hinges on a customer's choice of one firm's long-developed prototype over that of a competitor. At the time of the interviews at TELECO many engineers there were very discouraged by the fact that the P.T.T. had just rejected for further production a prototype on which some of them had been working for years. Some complained that the decision was political. All seemed to feel that the expertise they had acquired was suddenly less valuable, and that their careers would suffer.

Finally, if an engineer did not display any job insecurity, he was asked if he could imagine any conditions in which his job might be at risk. Forty-five percent of the PAMPCO engineers replied none, but that figure dropped to

29 percent at TELECO. Curiously, then, although more disappointed about their careers, the engineers in the depressed industry appear to feel slightly more secure about their present employment than do the engineers in the expanding, high technology industry. The replies to this last question also provided additional evidence of previously observed differences among categories of engineers. The *ingénieurs autodidactes* felt the most secure about their jobs, the *chefs de service* the least secure. The former seemed to assume that a company that had treated them exceptionally well, and which they had loyally served for many years, would be loathe to turn them out now. The latter seemed more aware than the other categories of the harsh economic realities confronting their companies.

Each engineer was also asked how easy or hard it would be to find another job if he were seeking one. Here the differences between the firms directly reflect the market situations of their industries. While only 23 percent of the TELECO engineers thought it would be difficult to find a decent job, over 50 percent of the PAMPCO interviewees thought it would be. The engineers at PAMPCO seemed to feel trapped in a stagnant industry, which is probably why only 47 percent of them said they plan to stay in their present job because they like it, as opposed to 86 percent at TELECO. Those who worked in the PAMPCO factory were most pessimistic, those at TELECO's Simulator Division the least so. As to the different categories, it was the *ingénieurs autodidactes* who were most pessimistic, with 60 percent saying it would be tough to find another job, as opposed to only 24 percent of the *ingénieurs diplomés* and 7 percent (one person) of the *cadres administratifs*.[15]

In summary, the engineers at PAMPCO displayed surprisingly little concern about keeping their present jobs, but considerable frustration about being stuck in them for reasons of their weak position on the external labor market. This was also true of the *ingénieurs autodidactes* at both firms. Those PAMPCO engineers who did acknowledge any insecurities seemed nostalgic for the paternalism of an earlier time that protected *cadres* against layoffs, and to resent becoming dispensable (like other workers), and the decline in their status that implied. The engineers at TELECO displayed a surprising degree of job insecurity, but much greater confidence that they could find something else if necessary. The sources of their insecurity were not the economic conditions of the industry itself, but rather the market and technical uncertainties within it and their doubts about the TELECO management's ability to provide far-sighted leadership. Such uncertainties, combined with their lower expectations for spending their entire career at the same company, and with their considerable confidence about finding another job, may well have contributed to their slightly more casual attitudes towards their careers. It remains to be seen whether this labor market situation helps account for any of their industrial or political attitudes.

Attitudes towards salaries

The previous section showed that engineering salaries at TELECO were somewhat lower than those at PAMPCO, but two questions remain. Were the same disparities found between the two subsamples interviewed, and how did they feel about their levels of compensation?

Not surprisingly, the average salaries reported by the interviewees at TELECO were slightly lower than those of their counterparts at PAMPCO. For example the average (mean) age of the twelve *ingénieurs diplomés* who were not *chefs* at TELECO's Simulator Division was 31, the average rank was 3.9, and their annual salary averaged 71,000 FF.[16] At PAMPCO's Technical Services Division, the nine *ingénieurs diplomés* interviewed had an average age of 32, an average rank of 3.8, but an average salary of 72,000 FF. Among the *techniciens*, the discrepancies were a good deal larger. For the four at PAMPCO's Technical Services Division, the average age was 29.5, the average rank was 2, the salary average 52,000 FF. For their four counterparts at the Simulator Division, the average age was 28, the average rank 2, but the average salary only 46,000 FF.[17]

Although the salary differences were quite small, the TELECO engineers were considerably less satisfied with theirs. Asked what they thought of the salaries paid engineers at their firm, 63 percent of PAMPCO interviewees, but only 36 percent of those at TELECO, said they were satisfactory. Seventy-two percent of all the *techniciens* were dissatisfied.

The discontent with salaries at TELECO appeared in responses to other questions as well. When those who said they sometimes wished they had gone into a different occupation were asked why, 24 percent at TELECO as opposed to 13 percent at PAMPCO emphasized better pay. Similarly, when asked if there were any "morale problems" at their firm, 25 percent at TELECO – 32 percent at the Simulator Division – but only 11 percent at PAMPCO stress pay. Moreover, the discontent at TELECO was with more than just the salaries, it was with the whole package of extrinsic rewards, including vacations, holiday policy, the food in the company cafeterias, the lack of coffee machines around the offices, harsh pay deductions for being late to work (*techniciens* only), and a general insensitivity to the personal and family difficulties often created by travel assignments or sudden relocation of a whole department. Nowhere was this dissatisfaction clearer than in the answers to the question: "Do you think top management here does all it can for the employees?" Table 7.9 displays the coded responses at PAMPCO and TELECO.

The differences between the two firms here are striking; while 58 percent of the PAMPCO engineers rate their employer as good or above average, none of the TELECO interviewees agrees. The following comments provide a sense of both the character and depths of the discontents at TELECO.

Table 7.9. *Company efforts by firm*

	Superior	Average	Minimal	N
PAMPCO	58%	27%	15%	40
TELECO	0%	51%	49%	51
N	23	37	31	91

It's the minimum at TELECO. They do strictly what the law requires. On a scale of 0 to 10, I'd give them a 1, maybe. They could be more open to the problems posed by the personnel. [an *ingénieur diplomé*]

They do the minimum. TELECO used to be much more advanced, but no longer. They could do much more, for they are making plenty of money. [a *chef de service*]

TELECO could do much more. It's company policy to wait until there is some protest before yielding. It's hard for us at my level, between management and a fairly organized engineering staff. [a different *chef de service*]

The criticisms are not randomly distributed, however, but rather concentrated in particular departments and categories. Among the sales engineers, for example, the complaints were especially bitter and widespread.

No. Less than average. Here the most important thing is making money. They really don't give a damn about the employees. Take the problem caused by moving the Omega Department from here to [a town on the other side of Paris]. They are contemptuous of (*on se moque de*) the human problems that that causes. As for salaries, retirement, training program, they do the minimum. [*ingénieur diplomé*]

No, they do the minimum. For example, if a holiday falls on a Tuesday or Thursday, they don't give you Monday or Friday off the way most companies do. As for salaries, they prefer to hire less competent engineers and pay them less. But it's a dangerous policy, because many people are quitting for better paying jobs. They waste a lot of time hiring and losing experience. [an *ingénieur diplomé*]

I don't know. I don't have enough information to judge what the company *can* do. But if you compare it to other such companies, TELECO is clearly behind. For example, in its policies on vacations and transportation allowances. It does strictly what the law requires, not more, the minimum . . . On the other hand, they have never laid off anyone. [a *chef de service*]

At both firms, there were large differences among the different categories of engineers; at the extremes 65 percent of the *techniciens* said their company does far less than it could for them, while only 6 percent of the *ingénieurs autodidactes* agreed. Yet, the character of the negative responses of the *techniciens* is different at the two companies. The following comments come from *techniciens* at TELECO.

They do all they can to fill their pockets. As far as salaries and benefits, they do nothing, really nothing.

Not in my opinion. The company does what it can for the stockholders to make a profit. Salaries could be better, especially for the lower levels.

No. They don't hesitate to relocate an entire division if it suits them. I would rank them below average by comparison with other companies, towards the bottom.

It's a power relationship. The company would do the minimum, except that the unions and market force it to do a little more.

They do the minimum. They could do more, I'm sure. The salaries are too low, but worse yet are the differences among departments and even within them among people of the same level.

These comments reveal not only greater bitterness, but also a deeper cynicism about the motives of employers in general. The same sort of cynicism is evident in the following comments by PAMPCO *techniciens*.

No, but I don't know many companies that do. From the social and salary point of view, PAMPCO follows the rule book of the employers' association very strictly. They do the minimum.

It's the *patron*, so naturally it doesn't do more than it has to, or rather is forced to by the law and the unions.

Like other places, they do the minimum. There is no day care center here. Vacations are what is required by law. They could do more. The pay is terrible, especially when one considers the number of years of technical studies. Why should an *ingénieur diplomé* start at 5,000 FF per month, but a *technicien supérieur* start at only 3,800.

With this last comment, however, there emerges a second theme in the complaints of the *techniciens* at PAMPCO, a theme that is not at TELECO. Consider the following.

PAMPCO does nothing for *techniciens supérieurs*. It tries to block them, to keep them down. A beginning *ingénieur diplomé* from Metz [a *petite école*] starts with 30 percent more than a *technicien supérieur*, but after five years is earning 100 percent more. My only raises have been due to seniority.

Compared to workers and *ingénieurs*, we are not well paid. When you compare our education and responsibilities, it's not right. A *technicien supérieur* with a BTS has five years more education than a draftsman with a CAP [a vocational degree] but their salaries are about equal. Yet an *ingénieur* with two or three years more education earns twice the salary we do. Technical work is considered the "garbage can" of the French educational system. It's not sufficiently valued. There is an enormous division between *techniciens* and *ingénieurs*, which is not always justified. A *technicien supérieur* who does the same work as an *ingénieur* should be paid the same as a young *ingénieur*.

I consider myself badly paid, and accept it only because my wife works, and because I expect to go to engineering school soon. It's ridiculous that the salaries of *ingénieurs* can increase to such high levels during their careers. By contrast, the differences among the *techniciens* are minuscule. There should be more room to progress.

The gap between *ingénieurs* and *techniciens* is larger than that between *techniciens* and *ouvriers*, which is not right. They should raise us to make it more equitable. It's not the absolute amount that concerns me so much as the relative amount.

Here we see a textbook example of relative deprivation. The positive reference group for the *techniciens* is that of the *ingénieurs* with whom they work, the negative one is blue-collar workers. They deeply resent the *cadre* barrier to their upward mobility in either rank or salary, and also the way that barrier blurs social differences beneath it. Contrary to the expectations of new working class theory, these *techniciens* do not identify with workers; they feel superior to them and resent being treated like them. They take credentials seriously, aspire to higher status, and favor promotions on the basis of individual merit, not just seniority. Indeed, it is because of these individualistic and "bourgeois" orientations that they find the *cadre* barrier to upward mobility so frustrating.

Why is this sense of relative deprivation stronger at PAMPCO than at TELECO? It is impossible to be sure, but one major reason may well be the proportions and composition of the various strata in the industrial hierarchies in these two industries. At TELECO there are few blue-collar workers, especially at the Simulator Division, and they tend to be badly paid women. More importantly, the *ingénieurs* are not such a privileged elite at TELECO. The extent to which this reflects their large numbers as opposed to TELECO's specific managerial policies is impossible to say. However, it is interesting to note that the *ingénieurs* at PAMPCO also seemed concerned about their earnings relative to those below them, and about any erosion in traditional differentials. The following comments are illustrative.

Interviewer: Do you think top management here does all it can for the personnel? [Some of the remarks below are also drawn from answers to a question about salary differentials.]

Yes [emphatically], but at times it seems they do more for the workers than for us *cadres*, perhaps because the workers' unions are more militant. Our acquired advantages are diminishing. I blame the government's policies for reducing income differentials, however, not the company. It is important to maintain salary differentials in order to motivate hard work and reward responsibility. (a *chef de service*]

They do a lot more for workers than for *ingénieurs*. For example, for work hours and vacation days, the workers are better off. *Techniciens* and foremen who work Sundays are paid overtime; *ingénieurs* are not. [an *ingénieur diplomé*]

Salary levels here are correct, but among the *ingénieurs* of the same age and schooling the salaries are too similar. There are too few incentives for superior performance. And between *ingénieurs* and *techniciens*, the differences are small, too small at the beginning of the career. After all, *techniciens* can become *cadres*. [an *ingénieur diplomé*]

My promotion to *cadre* didn't bring me much of a raise, just 4 percent. It's wrong to be increasing the minimum wage without providing matching increases at other levels. I remember all too well what I – and my father before me – earned at the beginning of our careers. [an *ingénieur autodidacte*]

I am against egalitarianism and the tendencies in that direction. I blame Giscard. Sure, they should raise the minimum wage, but not hold down *cadres'* salaries at the same time. The problem is the Plan Barre incomes policy, which affects us salaried workers, but not the self-employed. [a *chef de service*]

"Status panic" may be too strong a term to characterize these defiant reactions to any diminution of salary differentials, but these *cadres* certainly do resent the narrowing of the gap between them and their social inferiors – often their subordinates at work as well. Unlike the *techniciens*, though, they do not invoke any rational standard of justice, but rather tradition and "human nature." Against that standard, they feel their traditional loyalty and quiescence is being abused. Yet, unlike the engineers at TELECO, they tend to blame the government, workers' unions, and demagogues, not their employers. Thus, it is only at PAMPCO that there were comments expressing gratitude for the way the employer had treated the interviewee:

They don't do the maximum, but they do better than most companies. They are especially advanced in their personnel policies. PAMPCO was one of the first companies to adopt flexitime. [a PAMPCO *ingénieur diplomé*]

They do more than necessary. For me, it's been a good company, and I'll be happy to finish my career here. Some smaller companies may pay better, but they offer less security. [an *ingénieur autodidacte*]

They are OK. They have to protect the business too. Of course there has been a tightening of the salary range, which they said was coming. But everyone knew the salary range was too big, and with the recession something had to give. Sure the unions have become stronger since being legalized [within the factory], but we shouldn't credit them for the changes that had to come. It was once more familial, more paternalistic; now it's more impersonal, but I can't say that that's necessarily worse. [an *ingénieur maison*]

My salary is correct. I earn as much or more as my classmates from Sup'Aéro [an *elite grande école*]. As for salary differentials, I have mixed feelings, for I hope to be up there in top management one day. [an *ingénieur diplomé*]

This last comments reflects the prospective view of a young *ingénieur diplomé* who expects to rise to the top himself, and so has a long-term interest in the maintenance of income differentials. The career expectations structured into his position condition his view of his current situation. Most of the other *cadres* favored maintenance of the differentials below their own level, but thought that the salaries of higher executives were often excessive. The most common element in these pro-company comments, however, is the category of the respondents. Two of the four are *ingénieurs autodidactes*. Given the starting points of their careers – skilled work in most cases, they have enjoyed impressive upward mobility, and are grateful to the employers who rewarded them despite their lack of formal credentials. The converse of the frustration among *techniciens* over the *cadre* barrier is the appreciation among those allowed through it. But the problem may be growing in that an increasing number of potential *cadres* begin their careers as *techniciens supérieurs*, not blue-collar workers, and so bump up against the *cadre* barrier much earlier in their careers. In trying to understand the attitudes of salaried workers, it is important to keep these labor supply factors in mind.

In summary, salaries are of considerable concern to the engineers at PAMPCO and TELECO, but the nature of those concerns varies by firm and category of engineer. The PAMPCO engineers are fairly satisfied that PAMPCO compensates them as well as or better than most companies, but many of the *ingénieurs* and *chefs de service* are upset about the government's program for increasing the wages of manual workers while limiting raises at the top of the salary hierarchy. In reaction, they defend traditional salary differentials and criticize those who would reduce them. Such attitudes are hardly the kind anticipated by most Marxian or neo-Marxian theorists; they reveal not a growing sense of identification with the working class, but rather deep concerns about staying above it. Yet neither do such attitudes suggest the defense of an occupation so much as of a multi-occupational stratum in the hierarchical world of corporate bureaucracies.

The anger of the TELECO engineers, however, as well as of the *techniciens* at both firms, does suggest a more class-conscious hostility to greedy employers. Still, without evidence of deskilling, it would be premature to interpret such attitudes as signs of proletarianization. What is clear is that the TELECO engineers perceive their employer as prospering, as nevertheless paying below average salaries, and as doing in general as little as possible for its employees. Of course, it is psychologically easier for them to feel and express hostility towards their employer, for promotion was more automatic than at PAMPCO, future employment at the company seen as less certain, and the external labor market viewed as stronger. Finally, it is evident that the *techniciens* at both firms are bitter about their salaries as well as their blocked mobility. Yet even as they display some hostility towards employers, they show no disposition to view themselves as workers and make common cause with them.

Conclusions

The evidence presented in this chapter raises serious problems for all three theories under consideration. Professionalization theory – or at least Freidson's version of it – assumes a high level of solidarity among "knowledge workers" as a result of common educational bakgrounds and occupational interests. In France, however, engineers enter their occupation by means of several very different routes, and *autodidactes* continue to represent a substantial minority within the occupation. Moreover, the training in engineering schools and the procedures for hiring young *ingénieurs diplomés* reinforce managerial rather than professional outlooks. More important, the variations in opportunities, according to educational credentials, render solidarity highly problematic among those occupying similar positions at any one time. The great expansion in the production of engineers by *petites écoles*, combined with importance of educational credentials for careers, appears to be intensifying this stratification within French engineering. In some occupations stratification is conducive to leadership and solidarity, but in this case it is not, for the elite engineers are on career tracks that lead out of engineering, and most engineers seem to define success in terms of the speed with which they advance up the bureaucratic hierarchy towards higher managerial positions within the firm.

The existence of staged careers and the engineers' interest in mobility also raise problems for proletarianization theory. The latter seeks explanations of attitudes towards work and politics in the content of jobs, ignoring the value of jobs as more or less favorable stepping stones to better jobs in the future. Moreover, in focusing on jobs, and socially locating persons in terms of them, it overlooks the ways that characteristics of the individuals filling jobs shape the opportunities associated with these jobs. In France the same job may be a stepping stone to quite different futures for an *ingénieur autotidacte* and a graduate of the Ecole Centrale.

The evidence on salaries, unemployment, and career ceilings is inconclusive. Even if the disproportionately growing number of *ingénieurs* graduating from *petites écoles* is destined for lower positions in the managerial hierarchy than those enjoyed by French *ingénieurs diplomés* in the past, their *grande école* colleagues are doing about as well as ever. True, there is greater competition than in the past from the graduates of elite business schools, but the fact that an increasing number of young men choose this variant of the *grande école* route to higher management cannot be regarded as proletarianization. In short, the fourfold expansion in the ranks of engineers may have broadened the base rather than the summit of this elite occupation, but not at the expense of the summit's contraction. On the contrary, this expansion has created many opportunities for inter- and intra-generational mobility, as more young people have found places in the rapidly expanding and less selective *petites écoles d'ingénieur*, and more workers and *techniciens* enjoyed promotions to the grade of *ingénieur autodidacte*. The only way that engineering in France may have become slightly "proletarianized" is with respect to job security. Yet, even here it is hard to separate the effects of industry's newness from those of its technology and the general massification of engineering staffs. In any case, *cadre* unemployment remains half that of blue-collar workers. The only reason that even this low level of unemployment has upset *cadres* is that unlike American engineers, engineers in paternalistic France have traditionally enjoyed employment security.

New working class theory anticipates greater hostility towards top management in the high technology sector, but for reasons of the work situation, not the market situation. This chapter has emphasized the significance of labor markets (including the way career opportunities soften dissatisfactions with current jobs), the importance of labor market issues in the grievances of engineers, and the differences in the labor market situation of PAMPCO and TELECO engineers.

Finally, this chapter has called attention to the distinctively national organization of technical careers in France, with the emphasis on socialization for work as an organization man, the *cadre* barrier, educational credentialism, the "closed promotion" system, and the low rates of job mobility. At the same time, this chapter has suggested that the *cadre* barrier is best understood as a line between trusted and untrusted workers. It is only in later chapters that the trust issue is addressed directly, but what is already evident, is that the kinds of careers that service class theory expects for trusted workers are available to *cadres* but not to *techniciens*. Whatever the explanation, it is clear that the *techniciens* in this sample occupy a very different position in the labor market than do other engineers. Yet, there are other variations among these engineers as well. The PAMPCO engineers are more disappointed with their careers and more pessimistic about the future. The TELECO engineers are underpaid, face employment insecurities despite a prospering industry and company, but enjoy a strong external market for their labor. The *ingénieurs autodidactes* face a weak labor market, but are satisfied with their pay and careers. It remains to be seen whether such variations better account for variations in industrial and political attitudes than do variations in the work they do or the knowledge they use in doing it. It is to these matters that we now turn.

8

Trade unions and professional associations

Abandoned by both sides [employers and workers] engineers discovered they were neither fish nor fowl, that they constituted a third party imperiled on two fronts, in the sad position of an iron caught between hammer and anvil. Georges Lamirand, *Le Rôle social de l'ingénieur*, 1954: 264, writing about the General Strike of 1936.

By refusing to form . . . a union for *cadres* themselves, and by affiliating with a workers' confederation – no matter which one – the engineers demonstrated their feeling not of belonging to a "new middle class" as has often been said of them, but rather of identifying socially and economically with the working class. Serge Mallet, *La Nouvelle Classe ouvrière*, 1969: 199

The proportion of French engineers belonging to interest group organizations is almost as high as the proportion of French manual workers. But in contrast to both the industrial unionism of manual and lower white-collar workers in France, and the occupational organizations of craft workers and salaried professionals in the United States, the unionism of French engineers is the unionism of *cadres*, a multi-occupational stratum of trusted workers holding positions of responsibility within organizations of all types. And although engaging in the limited collective bargaining that has gone on in France, and occasionally calling strikes, the main *cadre* union in the private sector – the Confédération générale des cadres (CGC) has distinguished itself from other French unions by its defense of the status quo, its aloofness from political issues, and its "responsible" tactics.

All three major theories of social change in advanced industrial society suggest departures from such *cadre* unionism among French engineers. The French new working class theorists are the most explicit. Mallet, quoted above, foresees a growing radicalization and politicization of French engineers, resulting in a growing proportion of them, especially in high technology industries, joining radical working-class unions of skilled workers, technicians and *cadres*. He anticipates a shift in particular towards the kind of unionism represented by the Confédération française démocratique du travail (CFDT), a militant unionism characterized by demands for workers' control (*autogestion*) and sharp reductions in salary differentials.

Proletarianization theories are less explicit about the implications of deskilling for collective organization, but suggest efforts to defend traditional autonomy and privileges by means of more aggressive collective bargaining. They also suggest a growing emphasis upon the shared concerns of "alienated" (or "instrumental") workers: salaries, hours worked, and protection against unemployment.

Professionalization theories offer a very different vision of collective organization among "knowledge-based workers." In the words of Eliot Freidson (1973a: 52):

> The sociological rather than merely technical or economic significance of a long period of training in putatively complex and abstract skills . . . lies in its tendency to develop institutionalized commitments on the part of those trained. Such trained workers are inclined to identify with their skill and with their fellows with the same training and skill. They are prone to develop . . . occupational solidarity . . . and associations.

By occupational associations, Freidson means professional associations, not occupational unions. Professional associations eschew collective bargaining; they protect the privileges and autonomy of their members by limiting the supply of certified practitioners and establishing legal monopolies by licensing procedures.

This chapter examines both the extent and form of collective organization among French engineers and the attitudes of PAMPCO and TELECO engineers regarding their group representation. It assesses the meaning of differences between the traditional and science sector engineers and their implications for theories of change in the collective identity, organization, goals and tactics of engineers. It begins with an historical sketch of engineering organizations in France, for only against this background is it possible to understand the institutional context in which French engineers pursue their interests.

Engineering organizations in France

Historical background

Before 1914, most French industrial firms were small, family-owned businesses employing very few engineers. These engineers worked in close collaboration with their owner-managers in settings in which there were few intermediate grades linking management and workers. This gulf between engineers and workers was reinforced by the revolutionary syndicalism of the workers' unions, a radical unionism that rejected the intermediate grades and appealed little to them anyway.[1]

To the limited extent that these intermediate strata formed associations at all, they formed local, noncombative, and multi-occupational societies, such as the Société amicale des chefs de service, contremaîtres, ingénieurs, agents de maîtrise et techniciens, organized in 1892. There were also the local chapters of the alumni associations of individual schools. In the case of the engineering schools, the alumni associations included many owners and directors of firms as well as practicing engineers.

To be sure, there were professional societies, and the oldest and largest of these, the Société des ingénieurs civils de France (SICF), did promote the career interests of practicing engineers. But in line with the technocratic orientations of the *centraliens* who dominated it, the SICF argued, according to Shinn (1980a: 195–196), that "management functions, when not combined with scientific expertise, constituted a parasitic drain on industrial growth . . . " In short, the SICF promoted a concept of the engineer not as an autonomous professional within the industrial bureaucracy, but as the most qualified manager of that bureaucracy. Only engineers could knowledgeably and credibly instruct, supervise, and "lead" the increasingly large, organized, and restless body of workers, as well as make informed decisions about investments in new industrial product lines, equipment, and processes. By the beginning of World War I, the SICF enjoyed a membership of over 6,000 *ingénieurs diplomés*, but the fact that among the members were independent industrialists as well as salaried engineers and managers inhibited the emergence of employee consciousness.

The period from the end of World War I to the Popular Front victory in 1936 saw considerable growth in the size and bureaucratic organization of French industrial firms and the first major efforts to organize salaried workers. In the face of a surge in the unionization of blue-collar workers after 1918, and of a major strike in 1919, the Union des syndicats d'ingénieurs français (USIF) emerged in 1919 as an organization of engineers in several industries. With the strike wave continuing, it participated in the founding of the Confédération des travailleurs intellectuels (CTI) in 1920. The CTI was officially recognized by the government in 1925; however, "because of its mixed composition (salaried and self-employed individuals)," it was short-lived (Grunberg and Mouriaux, 1979: 77). The Fédération des associations, sociétés et syndicats français d'ingénieurs appeared in 1928, and suffered similar difficulties. After "giving birth" to a union of salaried engineers in 1936, it dropped the word *syndicats* from its name, and retreated into coordinating the activities of numerous alumni associations and technical societies. A separate and at times radical union of technicians, and then technicians and office workers (USTEI), enjoyed considerable growth in the 1920s, but also declined during the Great Depression. Less radical and more successful was the Union sociale des ingénieurs catholiques (USIC), which grew from 1,000 members in 1919 to about 9,000 by 1935 (Descostes et al., 1985: 82). The USIC represented a "third way" ideology that emphasized the special role of engineers as neutral designers, organizers, and arbiters, but that accepted a capitalism that allowed engineers to rationalize production.

The decade beginning in 1936 was critical in the history of collective organization by French engineers, for it was this period that saw the emergence and consolidation of *cadre* unions. With the Popular Front electoral victory in 1936, the subsequent wave of sit-down strikes, and the government-sponsored agreements between organized owners and organized workers, salaried engineers increasingly identified their own interests as those of the embattled middle classes. Feeling squeezed between Capital and Labor, they began to seek vehicles for representing their separate interests in the emerging new bargaining forums. Several major engineering organizations appeared in 1936 or reorganized as autonomous unions of salaried engineers. In 1937, the three largest of these, the USIF, SPID, and SIS merged to form the 20,000-strong Fédération nationale des syndicats d'ingénieurs (FNSI).[2]

An effective "third force," however, would have to be even larger and more representative of the middle strata. Thus, as Luc Boltanski (1983: 24–25) explains:

A debate then opened about the extent to which the engineers should welcome further recruits from what were often called the "sound elements" of the firm, which meant, simultaneously, all those who occupied positions of relative authority within their firm, i.e. foremen and supervisors, technicians, sales representatives, sales managers, accountants, etc. . . . in short, all those who . . . occupied positions and had dispositions which tended to align them with the positions and attitudes of the engineers. It was precisely this scattered aggregate lacking homogeneity, lacking organization, lacking identity and, so far, lacking a name, that began to be designated by the vague term "*cadre.*" The vagueness of the term was its great virtue, since it enabled one to avoid endlessly asking the question of who was inside and who outside the field of mobilization then forming around the engineers.

Although wanting to organize as large and heterogeneous a group as possible around them, the engineers worried that "the fusion of highly qualified engineers and unqualified '*cadres*' into a single organization entailed a risk of social pollution, loss of distinction, and, ultimately, disintegration" (Boltanski, 1983: 26). In 1937 the FNSI signed a collective manifesto with both the CTI and the newly organized Confédération générale des cadres de l'économie (CGCE) – a confederation of several white-collar unions and the first to use the name *cadres*. However, it maintained its separate identity. According to Boltanski (1983: 26), the engineers felt that a full merger would be acceptable only under the conditions that the lower level *cadres* "construct their social identity in their [the engineers'] image." In Boltanski's (1983: 26–27) words:

These conditions were fulfilled under the Vichy government. The Vichy adminis-tration created conditions which favored the objectification of this "new group," if only by giving legal existence to the *"cadre"* category. The Charte de Travail made the term *"cadre"* (which it borrowed from the Army) both official and widespread. In accordance with the ideology of the "third way," it institutionalized the Third Party, which was now embodied in the *comités sociaux* in the form of the "arbitrating third party" which brought together technicians, supervisors, engineers, administrative *"cadres,"* etc. . . . . The primary function of the tripartite committees was to out-number the representatives of the working class, who had one vote against one for the *"cadres"* and one for the employers.

Vichy dissolved the two large workers' confederations, the militant CGT and the Catholic CFTC, making it easier for the engineers and *cadres* to organize their own unions. The SPID, still a semi-autonomous union of *ingénieurs diplomés* within the FNSI, encouraged cooperation with Petain, and while regretting the suppression of the workers' confederations, applauded the new corporatism. After 1944, however, the autonomous engineering and *cadre* unions faced renewed competition from the large workers' confederations, as the latter recruited *"cadres"* into special sec-tions established for them. Thus, in October, 1944, the FNSI, the CGCE, and two other autonomous federations of technicians, supervisors, and lower level administrators merged to form the enduring and important Confédération génerale des cadres (CGC).[3]

Initially the workers' confederations (the CGT and the CFTC) opposed government recognition of the CGC, ostensibly because of the SPID's earlier cooperation with the Vichy government. The government delayed recognition on the grounds of the CGC being insufficiently representative. In response, the CGC organized a 24-hour strike on March 25, 1946. It won recognition in August and broader legal standing in additional decrees in 1947 (Grunberg and Mouriaux, 1979: 79–81). Thus, the main organization for representing the collective interests of engineers emerged as a union of *"cadres"*, an ill-defined and heterogeneous group of salaried workers that represented a defensive mobilization by the new middle strata in the context of the class struggles of the 1930s and the growing role of the state in indus-trial relations.

As the CFC consolidated its gains in the post-war period, the workers' confederations reorganized their *cadre* sections and continued, with varying success, to recruit engineers. The split of the CGT into the Communist-dominated CGT and the non-Communist CGT–Force Ouvrière (the FO) led many *cadres* to quit the CGT's post-liberation *cadre* union. In response, the CGT created its Union générale des ingénieurs et cadres (UGIC) which became the current UGICT in 1969, with the addition of *techniciens*. "Pro-fessing its autonomy from the CGT's manual and office workers' unions, the UGICT progressed among *cadres* to the point where even the CGC negotiated a common statement of grievances (*un accord revendicatif*) with it in 1974" (Grunberg and Mouriaux, 1979: 82). In the meantime, the FO had organized its Fédération nationale des ingénieurs et cadres (FNIC). "Thanks to the double affiliation of its *cadres* unions to industrial federations within the FO as well to their own 'categorical' federation, the FO succeeded in organizing a widely dispersed population" (Grunberg and Mouriaux, 1979: 83).[4] It should be noted that while both the CGC and the UGICT (CGT) define *cadres* in the broad sense to include technicians and foremen, the UIC (FO) and other *cadre* unions recruit only among *cadres* as they are defined in the *conventions collectives*, that is among *cadres* and *ingénieurs* in the strict sense that this study has observed.

During the 1960s the Catholic workers' confederation had also split over secularization. The result was a fairly small, traditional CFTC and a new bur rapidly growing and militantly radical Confédération française démocratique du travail (CFDT). Both established *cadre* sections, the CFTC its Union générale des ingénieurs, cadres et assimilés (UGICA), the CFDT its Union confédérale des cadres (UCC). However, the UCC was not an autonomous *cadre* federation within the larger "inter-categorical" confederation. Rather, CFDT *cadres* belonged to the same industrial unions as the blue-collar workers under them, and the UCC confined its role to research and publicity.

The CFDT's rejection of a separate federation for *cadres* and its emphasis on *autogestion* reflected and reinforced the ideas of certain French intellectuals about a "new working class." These positions also distinguished the CFDT's program from the programs of the other *cadre* unions, but did not prevent cooperation with these other unions. When the CGC concluded an agreement with the major employers' association in France concerning consultation (*concertation*) at the company level, the UGICT (CGT) abandoned its cooperation with it (under their *accord revendicatif* of 1974) and sought to forge links with the rapidly growing (especially among technicians) UCC (CFDT). In 1975, the UGICT and UCC published a joint declaration on the protection of *cadres* against unemployment. Nevertheless, the UCC continued to take a more radically egalitarian position on the much debated questions of salary differentials and the income ceiling for social security taxes.

These were precisely the issues on which the CGC had built its reputation as a defender of the separate interests of *cadres*. Throughout the 1950s and well into the 1960s the CGC promoted its well-known triad of *cadres*' concerns: (1) maintenance of the *hiérarchie salariale*; (2) preservation of the *cadres*' special *régime de retraite complémentaire*, a supplementary retirement fund related to the ceiling on social security taxes; and (3) reduced taxes. It is significant that all three of these demands concern the relative or absolute market position of *cadres*. The other major plank in the CGC platform, *concertation* (but not *autogestion*), has more to do with the maintenance of middle management's authority within the workplace, but here too there appears to be a longer range concern that democracy at work will not simply undermine authority, but rather that it will undermine the justification for *cadres'* higher salaries.

Unlike the CGT and the CFDT, the CGC also prided itself on being "apolitical" and "responsible." It rarely called demonstrations, even refusing to participate in the mass demonstrations of May 1958 in defense of the Fourth Republic. Illustrative of the few demonstrations it did call was one to defend *la liberté de salaires* in 1954, and some in defense of *la retraite complémentaire* in the 1960s and 1970s.

Yet it would be a mistake to regard the CGC as apolitical in practice. In the first place, its traditional triad of demands implies political stances in a country where such issues are largely determined by the central government. The political implications are clear in the following statement of the longtime leader of the CGC, André Malterre, as quoted by Groux (1985: 32): "The notion of hierarchy . . . is thus not simply an element of salary policy. It involves a whole concept of society; there is a choice to be made between a civilization of termites where egalitarianism serves to justify a collectivist and technocratic tyranny, and a humanistic civilization where political, economic, and trade union liberties are not condemned to disappear." Moreover, as Groux notes, the CGC addresses its demands to the state, not employers, for the most part.

More importantly, the dominant conservative faction within the CGC has maintained the "middle class" or "third force" outlook discussed above, an outlook "according to which *cadres* were defined more by their place in society and their social status than by their job, their place in the firm" (Benguigui and Monjardet, 1984: 114). Indicative of this middle-class mentality was the CGC's creation in 1977 of what they called *Groupes d'initiative et responsabilité*. According to Benguigui and Monjardet (1984: 114), these groups included officials from associations of professionals, peasants, craftsmen, and small business persons, and "rumour had it that this was a political initiative sponsored by the supporters of Jacques Chirac," the head of the major conservative party, the neo-Gaullist RPR. In short, the CGC was clearly involved in a form of class politics.

Beginning about 1965, however, the CGC's opposition to the government's moderate *contrats de progrès*, contracts that would raise the salaries of the lowest paid workers most rapidly, combined with the union's lack of militancy, generated some internal dissension. A few member unions withdrew (or were expelled) and formed an independent Union des cadres et techniciens (UCT) in 1969, a union that represents *cadres* in several of the large nationalized industries. In response, the controversial and autocratic leader of the CGC, André Malterre, initiated some reforms and organized a strike in 1970.

Nevertheless, the CGC did not achieve unity. In addition to the traditional conservatives, there were two other factions that Benguigui and Monjardet label "the modernists" and "the corporatists." The corporatists "saw themselves as industrial workers, to be sure, workers of a very special kind, but workers . . . and criticized the conservatives for being too conciliatory toward the government" (Benguigui and Monjardet, 1984: 114). As unemployment grew and purchasing power declined in the 1970s, the corporatists gained increasing influence within the CGC, and won control of it in 1979. Significantly, the leader of the corporatist faction, Marchelli, was the head of the federation representing *cadres* in the metal-working industry, including those at PAMPCO. It was the metal-working industry, not electronics, that faced high unemployment and stiff competition from foreign imports. Thus, the interests of metal-working *cadres* were as much threatened by President Giscard's efforts to deregulate the economy – the economically liberal Plan Barre – as by the Socialists' plan for focusing government aid on the industries of the future, the high technology sector. More in tune with their interests were the traditional protectionist policies of the Gaullists. Moreover, the metal industry *cadres* emphasize that the burden of their managerial responsibilities justifies higher salaries. The first engineer I interviewed at PAMPCO, a *chef de service* in production, said that although he voted for the majority in the recent election, he was really only voting *against* the left. He disliked the Plan Barre, and resented the government's plan for raising the salaries of the lowest paid workers "by fiat," while maintaining a ceiling on higher salaries, and thus "screwing us *cadres*; responsibility should be rewarded."

In 1980, the CGC organized an "Estates-General" of supervisory workers, and threatened to run its own presidential candidate. (The CGC has frequently taken a stand in presidential races, usually by threatening to oppose the conservative government in power for neglecting one of its natural constituencies.) Since the election of President Mitterrand, it has shifted from an initial stance of wait-and-see neutrality to active cooperation with the rightist opposition. Again, the issues in contention have been *cadre* authority within the firm and taxes. A Socialist bill proposing new rights for industrial workers was viewed as a threat to supervisory personnel. According to Benguigui and Monjardet (1984: 113), the Socialists' tax reforms were "so resented by all managerial personnel that even those who belong to the CFDT asked in February 1982 for a two-year moratorium on further tax increases . . . Meetings between the CGC and opposition political parties (RPR and UDF) resulted in communiqués vehemently critical of the left-wing government . . . "

An interview (Le Guen, 1982: 76) with Marchelli in 1982 captures much of the continuing anti-egalitarianism and status anxiety of the CGC:

Twenty or twenty-five years ago there were few *cadres*, and one saluted a *cadre* respectfully from six paces away. Today, their number has grown considerably, and due to the evolution of the organization of work and the increase in the general development of education, nothing any longer distinguishes *cadres* from other categories, not even on a financial level . . . We have adapted well to these changes, accepted seeing our salaries progressively caught up with by other social categories. The salary hierarchy is no longer 1:6, but a hierarchy of 1:2.5 is not acceptable . . . The last few years have pushed things too far.

From this historical introduction five significant features of engineering unionism in France emerge. First, French engineers are organized as *cadres*, not engineers, and this multi-occupational form of collective organization represents an evolution away from the more occupationally grounded engineering associations of the nineteenth century and the engineer-dominated *cadre* organizations of the late 1930s. In other words, there is no evidence here to support the expectation of growing occupational solidarities in the emerging knowledge-based society.

Second, the unionism of French engineers, like the unionism of French manual workers, is characterized by "union pluralism." That is, among engineers in a single laboratory or plant, some will belong to one *cadre* union, some to another, and some to none at all. There is no single professional association comparable to the American Bar Association, nor are there "closed shop" laws to prevent competing unions from dividing the employees at the plant level. Third, this pluralism reflects different ideological rather than regional, industrial, technical, educational or hierarchical differences among French engineers.

Fourth, French *cadre* unionism appears to be unique. In pre-World War II Germany, "[s]alaried employees saw themselves as members of a middle stratum that superseded occupational differences," but the horizontal line separating them from workers was drawn lower in the hierarchy – between manual and white-collar workers rather than *cadres* and non-*cadres*; also, the line in Germany was less compromised by other divisions, and more deeply institutionalized in law (Kocka, 1980: 26, 272).[5] In Italy, there is a persistent pattern of ideological pluralism within the union movement, but nothing comparable to the *cadre* unionism of France. France appears to be unique in its combination of stratum-based organizations of engineers and a union pluralism based on widely differing social and political ideologies.

Finally, the objectives emphasized by the *cadre* unions suggest that *cadres* are more concerned about their relative market situation than their work situation. The one work situation issue that arouses the CGC from time to time is that of *cadre* authority in the firm, an issue that seems unrelated to deskilling, and that evokes militancy not among technical workers in new industry, but rather among supervisors in the labor intensive metal-working industry.

Patterns of support for cadre unions

The question now arises: to which if any of these *cadre* unions do different categories of engineers and *cadres* adhere today? There are no entirely reliable figures for the general rate of *cadre* unionization, but according to Groux (1983: 90) the best estimate for 1981 is between 12.5 percent and 15.5 percent. This may seem like a small proportion, but it represents a substantial increase since the 1960s.[6] Moreover, it is rapidly approaching the 18–20 percent rate for all employed persons in France.

A recent survey by Grunberg and Mouriaux (1979) overstates the percentages of unionized *cadres* as a result of the over-representation of large firms in the sample, but is still of considerable value. This survey of almost 1,500 *cadres* yielded a unionization rate of 26 percent. It shows the rate to be much higher, however, in the nationalized and public sector, 43 percent, as compared to 22 percent in the private sector.

For the *ingénieurs* in Grunberg and Mouriaux's survey, the unionization rate rises to 29 percent, compared to about 25 percent for *cadres administratifs supérieurs* and about 23 percent for *techniciens*, but these are small differences.[7] More significant are the distributions by type of union; three quarters of the unionized engineers and well over half of the unionized *cadres administratifs supérieurs* belong to the CGC, but less than a third of the *techniciens* do. By contrast, about 40 percent of *techniciens* belong to unions affiliated with one of the two major workers' confederations, the Communist-dominated CGT and the radical CFDT, while only 7 percent of *ingénieurs* do (Grunberg and Mouriaux, 1979: 100, 101).

Grunberg and Mouriaux also present distributions by type of industry and other variables. Unfortunately, these are for all *cadres* rather than for *ingénieurs* or *techniciens* alone, but they are still revealing. Twenty-eight percent of *cadres* in the *métallurgie* industrial group belong to unions compared to 15 percent of *cadres* in *électronique*. Among the unionized *cadres* in *métallurgie*, 76 percent are affiliated with the CGC and 12 percent with either the CGT or CFDT, while in *électronique* 54 percent are affiliated with the CGC and 33 percent with either the CGT or CFDT. What emerges most clearly from these figures, however, is not a large swing towards left-wing unions among *cadres* in *électronique*, for the total percentage of *électronique cadres* in the CGT or CFDT is only 5 as opposed to 3 percent among the *métallurgie cadres*. Rather, far fewer *cadres* in *électronique* – 8 percent as opposed to 19 percent in *métallurgie* – belong to the CGC, and this loss has not been matched by gains among the competing unions nearly so much as it has been by decreases in the ranks of the unionized.

Table 8.1. *Works Committee elections, 1966–67 and 1979–80*

Union	1966–67 %	1979–80 %	College II[a] %	College III (cadres) %
CGC	21.6	23.1	15.8	38.3
CGT	19.5	17.6	24.5	6.8
CFDT	18.8	18.2	27.0	13.5
CFTC	2.9	(4)	4.8	4.2
FO	7.8	(10)	12.3	9.7

[a] *Maîtrise* and *techniciens* alone.

Although the vast majority of French engineers in the private sector do not belong to unions, about two-thirds of them do vote for representatives to the plant level Works Committees (*comités d'entreprise*).[8] The major activities of the *comités d'entreprise* concern the arrangement of company-sponsored recreation and "social welfare," but by law the *comités* are also to be consulted by management on all questions of factory organization. Thus, while far from powerful, they are the main forum for regular plant-level discussions between management and representatives of the employees. Since most candidates for the biannual elections to them run on union slates in three separate "colleges," the first for manual and office workers, the second for foremen and technicians, and the third for *cadres*, the results of elections to the *comités d'entreprise* often reveal more about union attitudes than do membership figures.

According to Groux (1983: 92), the results of the 1966–67 and 1979–80 elections for delegates to the *comités d'entreprise* were as shown in Table 8.1.

These figures show the CGC gaining support, and the more class-conscious worker confederations, the Communist-dominated CGT and the radical CFDT losing ground since 1966. More recent company elections for *prud'hommes* (conciliation boards) reveal similar trends continuing into the 1980s; the percentage voting for CGC candidates in Collège III increased from 36 percent in 1979 to 41.5 percent in 1982 (Descostes et al., 1984: 246).Such a pattern of change must be regarded as evidence against the new working class thesis. Noteworthy also in this table are the differences between Collèges II and III. While over half the *maîtrise* and *techniciens* vote for CGT or CFDT candidates, just 20 percent of *cadres* do. While nearly two-fifths of *cadres* vote for CGC candidates, only 16 percent of *maîtrises* and *techniciens* do. Here is a dramatic expression of the boundary between *cadres* and *techniciens*.

In their survey Grunberg and Mouriaux group the CGT and CFDT

Table 8.2. *Works Committee elections by status and industry*

	CGC %	Inter categorical %	Other unions %	Abstained %	Did not answer %	Total %
Techniciens	12	21	7	52	8	100
Cadres administratifs supérieurs	23	12	11	48	6	100
Ingénieurs	36	12	12	31	9	100
Métallurgie	33	10	8	43	5	99
Electronique	27	20	7	42	5	101
Chimie	31	12	5	41	12	101

Source: Grunberg and Mouriaux (1979: 112–13).

together as "inter-categorical" confederations. One useful way in which they analyze the responses concerning Works Committee elections is by industry. Table 8.2 presents these data for metal-working and electronics.

The most striking aspect of this table is the sharp contrast between the distributions for the *techniciens* and the *ingénieurs*. The majority of *techniciens* do not vote. Among those *techniciens* who do vote, almost twice as many support candidates fielded by one of the major workers' confederations as candidates sponsored by the Confédération générale des cadres. But three times as many *ingénieurs* vote for CGC candidates as vote for candidates sponsored by a workers' confederation.[9]

The distribution for *ingénieurs* is much closer to that of *cadres administratifs supérieurs* – administrative (including financial, commercial and legal) *cadres* who share the middle management ranks with *ingénieurs*, come from similar social and educational backgrounds, and enjoy similar chances for career advancement – than to that of *techniciens*. It appears that doing technical work – whatever the inherent contradictions in it – is not by itself a source of radical unionism. There is a good deal of evidence in the French sociological literature (C. and M. Durand, 1971; M. Durand, 1972; Sainsaulieu, 1972) of a much greater militancy and a somewhat greater radicalism among French *techniciens* than *ingénieurs*. The most common explanation for the militancy of *techniciens* is the blocked mobility of educated workers, especially of those who do much the same work as *ingénieurs*, are in daily contact with them, and thus are inclined to take them as a reference group and feel relatively deprived. I will return to explanations of these patterns of union attitudes after presenting my own findings, but it is helpful to note now the evidence at the level of national sample surveys of broad support for "responsible" *cadre* unionism among the *ingénieurs*, even in elections for delegates to *comités d'entreprise*, and the contrasting support for delegates associated with militant workers' unions among the *techniciens*.

As for the differences among industrial groups in Table 8.2, the similar distributions between metallurgy and chemicals suggest that the particularly high support for "inter-categorical" (CGT and CFDT) candidates in electronics represents something other than the advanced character of its production technology. It may well represent a considerably higher ratio of *techniciens* to *ingénieurs* in *électronique* than in *métallurgie*. Since the differences among French industrial groups are correlated with a number of other dimensions that affect unionism rates and industrial attitudes, including size of plant and management's personnel policies (integrative–paternalistic vs. impersonal), it is impossible to draw further conclusions from Grunberg and Mouriaux's figures.

To understand the meaning of these patterns of unionism and union support, it is helpful to examine the *attitudes* of engineers towards collective representation. This is not to argue that aggregations of individual attitudes alone determine the degree and form of collective organization, any more than they determine the amounts and nature of riots and revolutions. As Tilly (1978) and others have shown in recent studies of collective action, capabilities (especially organizational capacity) and opportunities are of crucial importance to the mobilization of individuals. In the case of unionization and union activity, capabilities and opportunities are shaped by a number of situational factors, especially the laws governing industrial relations and the state of the labor market. Nevertheless, individual attitudes are important, especially in a situation of open shops and union pluralism such as found in France. Moreover, at the heart of the debate about post-industrial society in general and the new working class in particular are arguments about the effects of changes in work organization on individual attitudes.

The engineers at PAMPCO and TELECO

Union membership and electoral support

Each engineer was asked whether or not he belonged to a union, and if so which one? Only one-quarter of the 104 (excluding TELECO sales engineers) who replied reported belonging to a union at present. Table 8.3 displays the distributions by firm.

Many more PAMPCO than TELECO engineers belong and have belonged to a union. The two most common explanations for not belonging to a union were that "it's not in my interest" (28 percent) (the percentage rose from 10 percent for *techniciens* to 50 percent for *chefs*) and that none of the available unions is suitable, either because it does not represent *techniciens* (eight of the eleven *techniciens* at PAMPCO), is ineffective (15 per-

Table 8.3. *Unionization by firm*

	Never belonged, never will	Never have, but could	Belonged once, but no longer	Belong now	Total
PAMPCO	4%	31	23	42	100 (N=52)
TELECO	44%	42	4	10	100 (N=52)
Totals	24%	36.5	13.5	26	100 (N=104)

cent of all respondents), or is too conservative, radical, or (especially) political (22 percent at PAMPCO, 35 percent at TELECO). Only four respondents (one *chef* and one *ingénieur autodidacte* at each firm) said they were against unions for engineers or *cadres* in principle. The most common complaint about the unions was that they were all too political (*politisé*), or that the CGC was too inclined towards corporatism and the others were too political. The percentages saying that the unions were too political and/or that joining one might "damage my career" were considerably higher at TELECO than PAMPCO, while many more (25 percent vs. 8 percent) of the PAMPCO engineers excused their non-participation on the ground that the only union they liked was ineffective.

These variations become more intelligible in light of the distributions of preferences for particular unions at the two firms. The vast majority (91 percent) of the unionized and previously unionized engineers at PAMPCO belonged to the more conservative and "categorical" CGC; at TELECO by contrast, the few unionized engineers were split almost evenly between the CGC and the more radical workers' confederation, the CFDT. The pattern here resembles that found by Grunberg and Mouriaux, except that the inter-firm difference is even more pronounced. In both samples many more traditional industry engineers belonged to a union, and the union favored by them was the "categorical" CGC.

However, this difference between PAMPCO and TELECO would probably be much less pronounced if the engineers at PAMPCO had more choice. Of the twenty-two PAMPCO engineers currently belonging to a union, only nine report that they belong to their particular union – the CGC in every case – because they agree with its program and tactics; the remaining thirteen claimed they belonged to the CGC because "it's the only one available," it's the union "all the *cadres* here belong" to, and the union in which the factory director encourages membership. There's no suggestion that many of these would have joined an inter-categorical union, had one been available. But it does seem true that many were dissatisfied with the

CGC and would have preferred another union, whether one currently existing (including the FO and CFTC) or an ideal one imagined by the engineer. By contrast, not one union member at TELECO said he belonged to his particular union because it was the only one available.

It is useful to consider related data from company elections for delegates to the *comités d'entreprises*. As indicated earlier, many observers consider company election results a better measure of union sentiments than membership figures. The ballots are secret, most *cadres* vote, and the voter often has a wide choice of candidates. Unfortunately, but significantly, the choice at PAMPCO was quite limited. The only union able to get *cadres* to run with its sponsorship was the CGC. Consequently, there were enormous company differences; at PAMPCO no *ingénieurs* voted for candidates representing inter-categorical unions (and thirty-three supported CGC candidates), while at TELECO twice as many electors (21) supported such candidates as voted for CGC-sponsored candidates (9). The concluding section discusses the meaning of these findings.

Professional associations

Although by far the most important, unions are not the only organizations through which French engineers can pursue their common interests. There are a number of professional and scientific societies that encourage participation by industrial engineers, but only 10 percent of the engineers interviewed belonged to one. Moreover, the distributions of these by *cadre* status and firm offer no support to theories linking professionalization to the greater use of theoretical knowledge and school-trained workers in science-based industries.

Thirteen percent of PAMPCO *ingénieurs* belonged to a professional society (excluding alumni associations); only 8 percent of TELECO *ingénieurs* did.[10] It may well be that the higher proportion at PAMPCO reflects nothing more than the stronger pressure from a paternalistic director to represent the company appropriately, but the low figure at TELECO reflects a real lack of interest there. The one TELECO *ingénieur diplomé* here who reported belonging to a professional society, the Société française des électriciens et électroniciens, said that membership was automatic for graduates of his *grande école*, but that he also paid dues in order to get the technical journal. Even among those who had published articles, membership in professional societies was rare and of little significance. At neither firm did any *ingénieur diplomé* express a view of his professional society as a group to which he looked for representation or protection of himself or his occupation.

Ironically, the only group that yielded a considerable number of positive replies to the question on professional associations was the *ingénieurs auto-*

didactes. Twenty percent of them claimed membership in a professional society, in contrast to less than 7 percent of *ingénieurs diplomés.* However, the reasons most belong are either social and/or educational. These non-diploma engineers often look to such associations for assistance – through symposia, conferences, journals, etc. – in improving their technical skills; they may also see them as ways of demonstrating to their superiors their seriousness and competence. Finally, the *ingénieurs autodidactes* are often somewhat isolated within their companies. At the same time they lack alumni associations to give them some sense of connection with fellow engineers outside their companies, if only for the pleasure of socializing together. Thus the main reason one *ingénieur autodidacte* at PAMPCO belonged to an association of metal-workers (*fondeurs*) was to go on periodic factory tours with other members, after which they all dined together.

Many of the *ingénieurs* however belonged to the alumni association of their engineering school: 61 percent of *ingénieurs diplomés* and 72 percent of *chefs de service.*[11] These percentages were somewhat higher at PAMPCO, probably because management there encouraged participation. At both firms many engineers laughed about these memberships, saying they were good for little more than an annual dinner and an occasional newsletter. Still, a quarter of them admitted to being more active members or holding offices.

Moreover, membership involvement is not always a good measure of an organization's significance. It certainly is not in the case of the alumni associations of French engineering schools. These associations constitute the backbone of the fairly powerful Fédération des associations et sociétés françaises d'ingénieurs diplomés (FASFID), itself the main group within the Conseil national des ingénieurs français (CNIF).

FASFID obtained its greatest success with the passing of the law of July 10, 1934, the law that regulates the use of the title of *ingénieur diplomé.* During the depression of the 1930s, engineering salaries declined and unemployment rose to an estimated 13 percent by 1932 (Groux, 1983: 18). Engineers complained that the more than doubling in the number of engineering schools since 1900 – from 41 to 86 – had resulted in an over-production of under-qualified *ingénieurs diplomés,* thus undermining the prestige and value of an engineering diploma. Through their alumni and other associations, they succeeded in obtaining legal protection for the title, and setting up a commission responsible for overseeing the educational programs of all engineering schools and for approving the establishment of new schoole (Grelon, 1986). This tripartite commission is composed of representatives of (1) the *grandes écoles* (and Ministries overseeing them), (2) employers, and (3) unions and professional societies of engineers (Ribeill, 1984). It quickly made its influence felt. While France had

Table 8.4. *Responses to question on need for collective representation*

	Emphasize national representation and local communications	Emphasize local bargaining with top management	*Cadres* don't need collective organization	Totals
PAMPCO				
Techniciens	2	9	0	11
Ingénieurs autodidactes	5	2	2	9
Ingénieurs diplomés	7	10	0	17
Chefs de service	7	1	2	10
Totals	21 (45%)	22 (47%)	4 (8%)	47 (100%)
TELECO				
Techniciens	1	5	1	7
Ingénieurs autodidactes	2	6	2	10
Ingénieurs diplomés	7	16	0	23
Chefs de service	5	4	0	9
Totals	15 (31%)	31 (63%)	3 (6%)	49 (100%)

progressed from producing 1,000 *ingénieurs diplomés* per year in 1900 to 4,000 per year in 1920, it was producing only 2,000 per year in 1940.

FASFID works closely with another influential association, the National Conference of Grandes Ecoles. The two together have long proved adept at resisting efforts to expand or democratize the *grandes écoles*. The latest attempt to expand the engineering schools came in 1984. The Socialists aimed to double the number of engineering diplomas being awarded by 1995. So well organized were the elite engineering schools and alumni associations that within forty-eight hours of the proposal, the Minister of National Education had received thousands of protest letters. It is hard to gauge the effect of such opposition, but it does seem that the most elite engineering schools are succeeding at maintaining their traditional class size.

It would be a mistake to compare these professional associations of French engineers to the American Medical Association or the American Bar Association. French *ingénieurs diplomés* lack the important powers of professional licensing, and must continue to compete with the *ingénieurs autodidactes* at work. On the other hand, among engineers in the industrialized societies, the French *ingénieurs diplomés* have been unusually successful. It is no accident that France has the smallest number of graduate engineers relative to its work force, or the highest salaries.

The need for organization

From the low rates of participation in unions and professional associations, it might be thought that most of the engineers at PAMPCO and especially at TELECO did not feel any need for collective organization. This is not the case, and the answers to a question about such a need throw considerable light on the larger debate.

Each engineer was asked whether *cadres* (or *techniciens*) "needed to be organized" and why. If they answered in the affirmative, they were asked their views of the proper program and tactics of an organization representing them (question 40). Table 8.4 presents the three most common types of replies.

One feature of this table that warrants notice is the tiny percentage of engineers in either firm who feel that they do not need an organization to represent them: 8 percent at PAMPCO and 6 percent at TELECO. A second thing worth noticing is the variation by status. Because the numbers are small, these figures on intra-firm differences must be interpreted with caution. Nevertheless, they are suggestive. Of the seven engineers who said *cadres* do not need organizations to represent them, four are *ingénieurs autodidactes*, engineers who have succeeded in rising through the ranks on the strength of individual promotions.

With respect to the reasons given for needing collective organization, the *techniciens* in both firms insisted that collective organization is needed to help them bargain more effectively with company management. By contrast, the *chefs de service* emphasized the usefulness of organizations for dealing with threats from workers and the state, as well as, or even more than, from company management. The strongest statement of an anti-worker position was made by a *chef de service* in production at PAMPCO: "*cadres* should be nationally organized to confront the pressure of organized workers. My interests are very different than theirs. The CFDT and CGT are my enemies." Several other *chefs* at PAMPCO, including one in the Centre de Recherches, noted that a *cadre* organization "can help in conflicts with subordinates." At both PAMPCO and TELECO a majority of *chefs* and a good many other *cadres* emphasized the role of a *cadre* organization in protecting *cadres'* salaries and pensions against new ceilings, taxes or other encroachments.

Cadres must be organized at the national level in order to fight leveling taxes and other political attacks on them. There is much less need for organization at the local level, for we are already close enough to top management . . . It's the leveling tendencies, and they are growing. In France you've got to organize to defend your interests. [*ingénieur diplomé*, TELECO]

Yes, it's very important to be organized. *Cadres* need a union to represent them to the public – they're too isolated – but not for collective bargaining. [PAMPCO *chef*]

Cadres should be as well organized as teachers in order to fight teachers better. [PAMPCO *ingénieur autodidacte*]

Such a view of a collective organization as useful for resisting attacks on *cadres'* privileges was widespread among *chefs* at both firms and all *cadres* at PAMPCO. It is not surprising that French *cadres* are especially concerned about public opinion and state action. The state in France determines many matters that in the United States are resolved by collective bargaining.

The differences by firm are particularly interesting. Among interviewees who were *cadres*, the TELECO engineers were more likely than their PAMPCO counterparts to say they needed an organization to represent them in negotiations with company management. While about equal numbers of PAMPCO engineers expressed the two different reasons for needing collective organization, twice as many TELECO engineers (31) emphasized collective bargaining with company management as emphasized national representation. Sometimes these replies implied solidarity with all workers.

It's necessary to be organized, but not as *cadres*, because when one forms such a class-based organization, it's at the expense of other workers. There are no reasons to distinguish the *cadres* from the rest of the personnel. [a TELECO *ingénieur diplomé*]

Yes, and as *salariés*, not just as *cadres*. [a TELECO *ingénieur diplomé*]

More often, however, they reflected a concern for *cadres* in particular, *cadre* salaries, *cadre* unemployment and *cadre* responsibilities.

Yes, to fight against lay-offs and for higher salaries. We are all in danger. The companies today care only about their profit margins. They no longer recognize loyal service. The *cadre* has become a *salarié* but still has different interests from workers. [a PAMPCO *ingénieur autodidacte*]

We need to be organized to protect ourselves, for we're being exploited (*exploité*) by top management. I'd prefer an autonomous local union rather than a big national one, but would consider joining a national affiliate if it were the only organization available and I liked its general orientation. Whatever organization it is, it should be for *cadres* alone and be a vehicle for participation in company management not just for disputes. [a TELECO *ingénieur diplomé*]

For a long time I felt that *cadres* didn't need to be organized, but the situation and I have changed. In the past the engineer was a bourgeois who was naturally and easily close to the *patron*. But with the creation of engineering schools for the masses, many engineers are *super-techniciens* who must be managed by another smaller corps of loyal managers, who are now the finance-types, the technocrats. In which case the engineers are obliged to defend their interests. And they aren't going to join workers' unions because they still dream of regaining their former status. They're going to join the CGC. [a TELECO *chef*]

A few *ingénieurs* indicated a preference for a union specific not only to *cadres*, but to *cadres moyens*. As one put it:

Cadres have their own problems, but there's still a need for more particular organizations – unions and "colleges" [for company elections] for people of my level, *cadres moyens*. A need to be organized among ourselves to represent ourselves within the company, and to defend ourselves in the face of a position, especially as regards salaries, that has been declining since 1968.

These comments express a common preference for specific collective bargaining. They also reveal subtle but significant inter-firm differences in the nature of the interests defended. It was pointed out earlier that *chefs de service*, especially those at PAMPCO, tended to emphasize the value of *cadre* organizations for maintaining authority over workers. This inter-firm difference exists as well among the lower ranks of *ingénieurs*. The following quotes are from PAMPCO *ingénieurs diplomés* and *ingénieurs autodidactes*.

Yes, because the workers are organized. The *cadres* should be strong *vis-à-vis* the workers. It is indispensable to be organized – now more than ever – in order to defend ourselves within the company. We *cadres* are always in the wrong in disputes with workers.

Even in speaking of salary grievances, the PAMPCO *ingénieurs* were more likely to emphasize their position relative to that of the workers.

Yes, because the others are organized. *Cadres* risk falling behind because the workers are unionized and catching up.

At TELECO, by contrast, the *ingénieurs* expressed considerably more concern about the decline in the status, pay, and responsibilities of an *ingénieur* relative to images of *ingénieurs* in the past. This is evident in the comments above and it is emphasized in the following remarks.

Yes, the problem is that the *métier* of an *ingénieur* is declining in value and is not protected. [a TELECO *ingénieur diplomé*]

Being more numerous than 30 years ago the notion of *cadre* has tended to lose value and depreciate . . . A union should work at reaffirming (*revaloriser*) the responsibilities and work of *cadres*. [a TELECO *ingénieur diplomé*]

I think that before long *cadres* will be semi-skilled workers (*ouvriers spécialisés*) and not *ingénieurs*. [a TELECO sales engineer]

The comments of these TELECO *ingénieurs* reflect the *petit ingénieur* phenomenon discussed earlier: the emergence in a vastly expanded occupation of a lower stratum of engineers without managerial responsibilities and without much voice in management decisions. Yet it should be kept in mind that their grievances are not new. Both Fridenson (1985) and Kolboom (1982) note the declining status and growing specialization of many French engineers in the first third of the twentieth century, as many of them become salaried technical employees in the newly emerging large corporations. These developments were one reason for the emergence of engineering unions in the 1920s.

The PAMPCO *ingénieurs* also complain about a decline in their status, about having become *salariés*, but the emphasis is upon the loss of their previous role as the strategic link – the *interlocuteurs* – between workers and management. Only since 1968 has management been legally obligated to deal directly with workers' union representatives at the plant level. Before that, the uninstitutionalized character of industrial conflict in France represented an exceptional situation for the advanced societies. Perhaps the special status of French engineers as privileged line officers reflected their value to management as trusted intermediaries between two hostile groups. If so, the establishment and growth of regular, direct communications between unions and management has deprived the French industrial *ingénieur* of one important basis of status and influence. Some of the *ingénieurs* who have suffered this decline blame management for not resisting the unions more effectively, but most blame the workers' unions and the governments that have responded to union pressures at the expense of the *cadres*. Thus, in the traditional, labour-intensive industries, the *ingénieurs* emphasize their need for *cadre* unions to defend them in the national arena as much as in their direct negotiations with management.

In the science-based industries the vast majority of engineers have little or no contact with manual workers, and are less conscious in general of workers as an issue. It is not the workers' unions that disturb them, but rather the unexpected distance between themselves and higher management, and, for the growing proportion of *petite école* graduates, the stratification within the ranks of *ingénieurs*. At TELECO's rapidly expanding Simulator Division, moreover, management's strategy for hiring engineers represented and reinforced a relationship of low commitment by both parties to a long-term relationship. Unlike the case at PAMPCO, where engineers were carefully selected for a lifetime of service, TELECO's engineers were hired hastily, weeded out again after two or so years, and badly paid in the meantime. Many of the young engineers displayed a similar lack of commitment to the company, regarding their job there as a temporary one to gain experience at a reputable firm.

Nevertheless, all who are *cadres* share a social status that defines them as having a good deal more in common with industrial managers than with manual and lower white-collar workers. And while this *cadre* status may mean little in the daily work of a young engineer in a high technology firm, it means a good deal educationally, socially, legally, and especially in terms of career prospects. As such, it continues to inhibit any strong identification with workers. The result is a tendency for science sector engineers to see themselves less as members of an embattled middle block than as individuals in an ill-defined and stratified middle category of salaried workers. As long as the existing unions continue to represent the embattled blocks – of *cadres* or all workers – of traditional French industry, these science sector engineers will shy away from active participation in any of them. At the same time, those at TELECO will continue to feel a need for stronger collective representation than offered by the CGC in intra-firm dealings with management, and so will give considerable electoral support to inter-categorical unions in company elections, unions whose ideologies however have little deep appeal. Many who vote for CFDT candidates do not favor *autogestion* or a reduction of the *cadre*–worker salary differentials. But they see their support for a union that favors such measures as their best available way for impressing upon management their concern for reducing the differences between *cadres moyens* and higher management in the areas of salaries and decision-making influence.

Conclusions

The findings examined in this chapter provide little support for either new working class or professionalization theory. With respect to the former, it is true that among the interviewed engineers, those in the high technology setting are more likely than their traditional industry counterparts to join or support a radical, inter-categorical union as opposed to an exclusively *cadre* one. On the other hand, this support appears to be more tactical than ideological – an expression of militancy in a market situation that offers fewer rewards for loyalty and fewer punishments for collective action against management than would be the case in a more paternalistic and economically depressed industry. It is clear from the evidence in chapter 6 that it does not represent support for *autogestion*, an official doctrine of the CFDT and a basic tenet of the new working class as conceived by Mallet.

Professionalization theory fares no better. Like other salaried workers in France, especially those in relatively privileged situations, French engineers identify with a stratum within the industrial bureaucracy rather than a specific occupation. This stratum may be the broad one of salaried workers at times, or more commonly the much narrower one of *cadres*, but it is rarely engineers. True, the engineering schools and their alumni associations have achieved some influence over the supply of *ingénieurs diplomés*. However, the consequence of this influence is limited by the fact that industry remains free to create *ingénieurs autodidactes*. Furthermore, the strength and role of alumni societies introduces a divisive element within the occupation. French *ingénieurs diplomés* identify and cooperate with fellow graduates of their school, whether or not they are practicing engineers, or with *cadres* in general; in the process *ingénieur* loses its significance as a meaningful category. Among French engineers in industry, there is nothing comparable to the occupational power of American physicians in hospitals. Moreover, with the rise of both the *petites écoles d'ingénieur* and the business schools, and the related decline in the prestige of the *ingénieur*, the trend is away from identification with the once proud category, *ingénieurs*, and towards identification with the multi-occupational category of *cadres*.

Proletarianization theory makes only a little more sense than the other theories. It correctly suggests a growing opposition to management in defense of traditional privileges. It is sensitive to the increased concerns of engineers, especially those in the science sector, about salaries and job security. However, proletarianization theory is wrong about the causes of these developments. Chapter 5 revealed no evidence of any deskilling of engineers, and showed how much such "proletarianization" is actually the result of an enormous expansion in the production of *petits ingénieurs*.

It is nonetheless true that the authority and market situations of the *petits ingénieurs* in high technology industries are conducive to a rejection of traditional *cadre* unionism. Engineers in the high technology sector have a much less problematic relationship with the working class than do their counterparts in traditional settings. This is because there are few blue-collar workers in the knowledge-intensive industries, and few engineers working in production. Moreover, the programs of the working-class parties pose less threat to the semi-public high technology sector than do the protectionist policies of the Gaullists or the free market policies of the other conservative party. At the same time, the CGC is enjoying fresh support within old industry because of the new threat there to *cadres*' special role as privileged *interlocuteurs* between workers and higher management.

These developments in the patterns of engineering unionism in France are substantial, and they are related to changes in production technology. But contrary to all the theories under consideration they reflect technology's effects on labor markets as well as work situations, markets whose structures bear the stamp of such particularly French institutions as the *grandes écoles*, *cadres*, and industrial paternalism. Finally, they reflect the distinctively French political and union contexts that condition the responses of French engineers to changes in their market situation.

To be sure, there is a common thread in accounts of the collective organization of engineers in advanced societies. In their studies of British and American engineers, both Whalley and Zussman find engineers struggling to protect their market position without compromising their beliefs and interests in individual career mobility. Whalley emphasizes the felt need for collective representation, but also the reservations about conventional collective bargaining and about the Labour Party politics of the more effective unions. Zussman finds greater militancy among the advanced sector engineers, and attributes it to the instability of their employment situations, many of them having experienced lay-offs and expecting to change companies again during their work lives.

These concerns are echoed in the statements of the French engineers, for the straightforward reason that in all three societies engineers are employees of large industrial corporations. As such, they are unable to establish the kinds of monopolies over the market for their services that physicians have achieved (Larson, 1977). And as trusted salaried workers, they enjoy career opportunities that make collective bargaining of limited interest.

Yet, these common concerns are expressed differently in the three societies. In Britain, there were few differences among engineers in the old and new industrial settings. In the United States, there were differences but not of the kind found in France. Whereas it was in the traditional French firm that many more engineers belonged to a union, it was in the science-based American one that engineers were most supportive of a union. Where the TELECO engineers showed a greater willingness to join with working-class unions, the advanced sector American engineers preferred a professionalization strategy. Such national variations are best understood by reference to the institutions governing the production, recruitment, employment, and promotion of engineers, and to the historical conditions that generated engineers' specific coping strategies.

9

Social participation, politics and class

The position of technical workers in capitalist society raises questions about the interests, allegiances, and political tendencies of what could be called the technical stratum of the "new middle class." Neither owners of productive property nor wage workers performing routinized labor, these educated employees enjoy moderately high incomes and considerable job autonomy, yet have little control over the products of their labor or the market for it. Thus, these traditionally quiescent salaried workers occupy an ambiguous class position that would seem to make them particularly susceptible to processes of industrial change. These observations give rise to three general questions addressed in this chapter: what similarities and differences are there in the class and politics of low and high technology engineers; what accounts for these patterns; and what are their implications for theories of class and social change?

Despite their differences, theories of proletarianization and a new working class concur in explaining the politics of technical employees by reference to their work situation. Nor is it only Marxists who tend to generalize from work place relations to class structure and national conflicts. Many liberal critics of Marx, most notably Blauner (1964), Lipset (1967), and Kerr et al. (1962), view the organization of work as the basis of the larger social cleavages in capitalist society.

Yet, a growing body of literature challenges such division-of-labor determinism. Some critics see a growing encapsulation of the once central conflict between labor and capital. Thus, for Daniel Bell (1973: 164) "[t]he crucial fact is that the 'labor issue' *qua* labor is no longer central, nor does it have the sociological and cultural weight to polarize all other issues along that axis." Other critics go farther, seeing life off the job as not only an autonomous source of politics, but as a source of industrial attitudes and behavior. In a critique of new working class theory, Low-Beer (1978: 230) writes that for Mallet and Gorz "[t]he causal links between life at work and life outside are all in one direction: from work to other realms. Our findings show, however, strong causal relationships only in the opposite direction: from attitudes acquired outside the work situation to militancy in the work situation." Sabel (1982: 190) emphasizes the "ambiguity of interests in relation to the division of labor" and the political significance of both previously acquired views of what constitutes an acceptable "career at work," and of particular work-place conflict experiences. Even such structural Marxists as Wright (1979), Carchedi (1977), and Crompton and Gubbay (1977) view the position of technical workers in the "relations of production" as so "contradictory" or "structurally ambiguous" as to render their class practices susceptible to outside influences.

To the extent that the work place does not account for political concerns and social solidarities, what does? Some observers (Alt, 1976; Kirchheimer, 1966; Lane, 1962) see the emergence of a "privatized" person who is socially isolated and politically passive. Others (Janowitz, 1978; Halle, 1984) call attention to such alternative sources of political concern and involvement as the family and the neighborhood. In his study of American engineers Zussman (1985) emphasizes the "politics of residence."

This chapter addresses these issues through an analysis of the social participation, politics, and class imagery of the engineers at PAMPCO and TELECO. In order to consider the political implications of these employees' non-work roles, the chapter begins with an examination of their involvement in families and local communities. Subsequent sections examine their political participation, voting behavior, political ideology, and class imagery. The final section evaluates the utility of the concept of a "service class" for understanding the position and politics of French engineers.

Work, family, and community

The engineers at PAMPCO and TELECO are not only technical workers; they are also husbands and fathers living with their wives and children. In a society that includes significant numbers of single workers and unconventional households, their marital situations are strikingly traditional. Among the interviewees, only two – one at each firm – are divorced, and only two – again one at each company – are "living with" rather than married to their domestic partners. The proportion of bachelors at TELECO is twice that at PAMPCO, but is still only 9 percent. Of the married men, all but one of the *ingénieurs autodidactes* and *chefs de service* have children, and except for the cases of three older *ingénieurs autodidactes* at PAMPCO whose children have grown up and left home, all have children living with them. Despite their younger age, the vast majority of *ingénieurs diplomés* and a substantial majority of *techniciens* also have children.

For all these categories of engineers, the families they live in are nuclear families in the sense that no other kin live under the same roof. Indeed, their relatives usually live a long distance away, a reflection of the fact that almost all the engineers live in a different town than the one in which they grew up. Even at PAMPCO, where 56 percent of the engineers are from the same department of the Lorraine as that in which they lived at the time of the interviews, only 10 percent live in their hometown.

Despite these similarities in family situations – similarities across companies and ranks – there are also significant differences. An important one involves the educational, occupational, and employment statuses of the wives. At both firms, the wives of the *techniciens* are far less educated than the wives of the *ingénieurs*. At PAMPCO, only 10 percent of the wives of the *ingénieurs diplomés* lack a high school degree (a *Bac* or *Bac Technique*), compared to 60 percent of the *techniciens*' wives. At TELECO, 54 percent of the wives of the *ingénieurs diplomés* have a university or *grande école* degree; none of the *techniciens*' wives do. The occupations of these wives reflect their educational levels. More than two-thirds of the wives of *techniciens* and *ingénieurs maison* work in clerical or blue-collar jobs; the figure drops to below 15 percent for *ingénieurs diplomés* and *chefs* at PAMPCO, but interestingly, only to 35 and 40 percent at PAMPCO. Not all these wives were employed at the time of the interviews, but here too there are characteristic differences between the firms and especially the ranks. Eighty and ninety percent of the wives of PAMPCO and TELECO *techniciens* are employed, but for the *ingénieurs diplomés* the figures drop to 50 and 57 percent respectively, many of whom (50 and 25 percent) worked part-time.

The relationship between family and work

The separation of work and family is often treated as a defining feature of modern industrial society, but it is easily exaggerated. Visitors to France are familiar with that country's delightful multitude of family-run bakeries, bistros, and similar small businesses. In these settings, work and family lives tend to be closely integrated; positions in the family determine work activities, and the demands of work shape the rhythms of family life. For independent professionals too, there is a strong if often more stressful relationship between these two domains, with the professional's involvements in work intruding in various ways on the lives of spouses and families.

By comparison, there is little relationship betwee the work and family lives of engineers. As Zussman (1985: 175) notes, "engineering employment – if not salaried employment in general – imposes few demands on the employee's life away from work, whether of time beyond a 'normal workday,' or life-style or of obligations from family members. This is the structural basis of a sharp differentiation of work from family." Yes, three aspects of engineering in France complicate this portrait. The fact that most engineers are *cadres* and many are middle level executives means that there are often managerial responsibilities that keep the engineer at work beyond the "normal workday." Thus, *chefs de service* worked longer hours than *ingénieurs diplomés*, and the latter worked longer hours than *techniciens*. The great majority of the engineers said they never or rarely took work home with them; yet, 24 percent of the *chefs* (and 20 percent of the *ingénieurs maison*) reported taking work home at least once a week.

A second aspect of French engineering, however, tends to limit the intrusion of the work of executives into family life. The fact that careers in France are more predetermined by educational credentials and seniority than in other countries means that managers feel less constrained to stay late at the office or call on their spouses for help – be it with typing, entertaining, or as community volunteers (Kanter, 1977; Mortimer and London, 1984). Whatever the reasons, the "two-person career" (Papanek, 1973) that is so widespread among single-provider managers and professionals in the United States appears to be far less common among French *cadres*.

The third aspect concerns the persistence of paternalism in many old and provincial companies. At PAMPCO, all ranks claimed to work longer hours than their TELECO counterparts, and to suffer far more family life disruptions in the name of emergencies at work. To some extent, such problems are inherent in semi-continuous production, especially in such breakdown-prone industries as traditional metal-working. This was one major reason that PAMPCO provided some company housing, and encouraged production and maintenance engineers to live nearby. In fact, 60 percent of PAMPCO engineers (vs. 20 percent at TELECO) lived less than ten minutes from work; 22 percent of them (vs. none at TELECO) lived in company housing just across the street from the factory. Whatever the justifications, the family lives of these engineers were all the more vulnerable to intrusions from the world of work. Moreover, these intrusions were not limited to demands upon their time. For the engineers living in PAMPCO's pretty hamlet of *cadre* housing, family life was literally visible to their work place superiors, and the engineers felt under pressure to maintain a respectable life-style. Nor was there any protection at the normative level. The factory director spoke frankly and unapologetically of his expectations for proper behavior off the job by "his" *cadres*, and expressed satisfaction with the way most were currently conducting themselves as parents and local citizens. He took pride in his ability to help them with domestic "problems," and few engineers questioned his right to do so.

This situation at PAMPCO's original factory may represent a somewhat extreme case of company paternalism in contemporary France. Nevertheless, it would be a mistake to imagine that such a residential integration of work and family life persists only in the most traditional settings. Electricité de France (EDF) is France's nationalized electrical utility, and its nuclear power plants represent a clear case of high technology production. In the course of some research I conducted at one of these nuclear power plants, I found that there too a great many engineers and their families lived in a self-contained company village during their two to four year tour of duty at that power plant.

For the engineers at the three TELECO sites, the situation was quite different. Only 18 percent lived in the same town (or *arrondissement* in the case of the Paris-based headquarters) as their work place; 21 percent lived more than thirty minutes away (vs. 2 percent at PAMPCO). The company made no efforts to influence their behavior off the job. One way, however, in which the demands of work impinged upon the family lives of some outside the normal hours of work was in the travel and relocation obligations associated with certain positions. Big, growing, multi-plant companies often wish to transfer certain types of staff members to other locations. Several of the engineers working at the telephone exchange components plant in Brittany had been working at the Paris headquarters a few months earlier. Some were happy to escape the capital, but others were not; several were commuting daily from Paris and uncertain about their plans.

In addition, companies engaged in making such complex items of equipment as telephone exchanges must provide technical personnel to install and service them. One engineer I interviewed had spent two years in Las Vegas refining and installing a sophisticated telephone exchange. Roughly half the TELECO *ingénieurs diplomés* (and one-third of those at PAMPCO) reported having had to move at least once for work-related reasons. This may not seem like a lot to mobile Americans, but in the context of French culture it creates significant problems of community integration. In the frustrated words of an EDF engineer who chose to live "in town" rather than the company *cité* so that he could participate in local life: "you have to live here for ten years before people stop treating you as a foreigner." For the most part, however, there was a firm structural basis at TELECO for the separation of work and family lives. At PAMPCO, there was a considerably weaker one.

Turning now to the subjective level, most of the engineers valued this separation and endeavored to preserve it. Two-thirds of them were emphatic about the importance of maintaining a "barrier" between the two spheres. Which sphere were they protecting from which, and how well were they succeeding? As Chapters 5 and 7 show, most of these men took their work and careers quite seriously. But valuing family life as they did – not as a support system for their work lives but rather as the source of their most meaningful satisfactions – they consciously limited their psychological involvement in their work roles. Their goals were to keep their work and private lives apart but harmoniously balanced, each complementing the other. Their fears were that work would interfere too much with life off the job. At PAMPCO it often did. Despite the greater willingness there to adapt private life to the demands of work, only 24 percent of the PAMPCO *ingénieurs diplomés* (compared to 64 percent at TELECO) expressed satisfaction with the balance they had established between the two spheres. The satisfaction figure is higher for *techniciens*, but drops to 12 percent for *chefs* at both firms. It is also low for the *ingénieurs autodidactes*, especially at PAMPCO. The main source of dissatisfaction was time. To a question about the main impact of work on family life, over half complained about the excessive time demands of work. In contrast to blue-collar workers (Rubin, 1976), very few mentioned money, either in a positive or negative way, and few mentioned the consequence of work on their moods off the job, the exception here being bad moods among the PAMPCO *techniciens*.

All this suggests that both those with managerial responsibilities and those whose careers depend on pleasing local superiors, find it difficult to limit the place of work in their lives. To the extent that high technology industry employs greater proportions of *ingénieurs diplomés* in non-managerial positions, there may be a trend towards a more successful differentiation of work and non-work roles among French engineers, as well as a widening gap between the private life-styles of *chefs de service* and *ingénieurs diplomés*, and a narrowing one between the life-styles of the latter and other salaried workers, *techniciens* in particular. Whether for reasons of self-selection or different company norms, the engineers at TELECO were more insistent about the sanctity of their private lives. Recall the angry demonstration by dozens of them outside company headquarters over the unilateral decision to transfer an entire department of the Simulator Division from one location to another. In short, the differentiation between the work and non-work lives of French engineers may not be as great as it is for American engineers, but it appears to be growing with the decline of their managerial responsibilities and of company paternalism. The question that remains concerns the alternative involvements, identities, and interests that may arise in the non-work sphere.

Residential and occupational communities

In his study of American engineers, Zussman (1985) observes that the residential dispersion of these salaried workers inhibits the formation of occupational or industrial communities. Yet, it does not result in "privatization," despite the fact that these white males lack the kinds of ethnic, religious, or minority statuses that often provide an axis in America for non-work-based solidarities. Rather, home ownership and the local character of such key family matters as neighborhood safety and public education conduce to the development of "residential communities." These territorial solidarities create bonds of shared interests and sentiments among individuals that occupy widely different positions in the division of labor, includ-

Table 9.1. *Non-work contacts with workmates and neighbors by rank and firm*

Table 9.1a: Percent knowing one or more neighbor

	Techniciens	Ingénieurs autodidactes	Ingénieurs diplomés	Chefs
PAMPCO	9 (1)	40 (4)	23 (5)	50 (5)
TELECO	31 (4)	55 (6)	27 (7)	73 (8)

Table 9.1b: Non-work contacts with workmates
(Percent saying see one or more workmate outside work)

	Techniciens	Ingénieurs autodidactes	Ingénieurs diplomés	Chefs
PAMPCO	58 (7)	56 (5)	79 (15)	60 (6)
TELECO	73 (8)	36 (4)	63 (15)	55 (6)

ing both blue- and white-collar workers. For evidence, Zussman points to the friendship patterns of his interviewees.

It is instructive to see what the same methodology reveals about the social bonds of the French engineers. Each interviewee was asked to briefly describe his two or three best friends, including their education, work, and residence. Additional questions concerned relations with neighbors and off-the-job contacts with work place colleagues. Table 9.1 displays the findings on relations with work associates and neighbors by rank and company.

The first thing to note is that in both companies over half the engineers say that relations with neighbors are either non-existent or limited to the formalities required by politeness. In some cases, the wives and children socialize with some neighbors, but only 33 percent of the PAMPCO and 43 percent of TELECO engineers report knowing one or more neighbors well. The figures for the *ingénieurs autodidactes* and the *chefs de service* are higher than for the other ranks, probably because these men are older and thus more likely to own their own homes and to have lived in the same place for several years. More surprisingly, the figures for PAMPCO are lower than those for TELECO. One reason may be that the PAMPCO engineers are busier seeing colleagues from work.

Over half (57 percent) of the TELECO *ingénieurs* report socializing with work associates off the job, and the figure rises to 70 percent for the PAMPCO *ingénieurs*. At this aggregate level of analysis it appears that work-based relationships do spill over into life off the job, especially at PAMPCO, while residentially based relations are relatively weak. One reason for the weakness of local bonds may be that few of the engineers at either firm grew up in the town where they now live; the figures are 10 percent at PAMPCO and 16 percent at TELECO. Moreover, 87 percent at

PAMPCO and 90 percent at TELECO have moved at least once, and more often two or three times, since taking their first job.

The data on friendships throw additional light on the sources and character of these engineers' social bonds. Table 9.2 looks at friendships in terms of where they originated, breaking the figures down by rank and company. What is most striking here is the predominance of work-related sources over neighborhood and voluntary associations: 30 percent of the friendships were formed at engineering school (or during the *cours préparatoires*); another 20 percent at work. By contrast, only 6 percent of friends were met as a result of sharing the same neighborhood or participating in the same voluntary association. After occupational schooling and work, the next largest source of friends is "youth," evidence perhaps of the French tendency to be slow at developing close friendships, but to regard them as enduring, once made.

Within each firm, the distributions are remarkably similar. Between the ranks, however, there is one important difference: while friendships formed in childhood are the most important for the *techniciens*, and friendships formed at work the most important for *ingénieurs autodidactes*, occupational schooling is the most important by far for the *ingénieurs diplomés* and the *chefs*. Like academics, graduate engineers in France undergo an extended period of training that is known for its selectivity and rigor. Many lasting friendships begin during the ordeal of the *cours préparatoires*. Moreover, the elite status of the most prestigious *grandes écoles* creates a predisposition towards relations of mutual aid among all graduates of one's own school. Nowhere is this more evident than in the tradition among *polytechniciens* of using the familiar form of address, regardless of differences in their ages, rank, or personal knowledge of one another. Evidence of a negative sort is the lower schooling figure in Table 9.2 for *cadres administratifs*, for one of them had no higher education, and four had attended a university rather than a *grande école*.

At least as important as the origins of friendships is the occupational status of the friends. Working in the same company or industry, but not necessarily the same occupation, suggests the existence of an industrial community, the type often found among miners (Gouldner, 1954) and steel workers (Kornblum, 1974). If friends work in the same occupation, that may indicate the kind of occupational community frequently found among craftsmen (Lipset et al., 1956) and professionals. Table 9.3 presents findings on some relevant characteristics of the engineers' friends.

Noteworthy here is the higher percentage of PAMPCO respondents who report that one or more of their best friends works at the same firm as they do. Even more striking are the high percentages of *ingénieurs* who report that most or all of their friends are in the same occupation as themselves. This is not surprising, given that many of the higher status engineers met their friends in professional schools, but since it includes the *ingénieurs*

Table 9.2. *Friendships by origins, rank, and firm*

	Techniciens	Ingénieurs autodidactes	PAMPCO Ingénieurs diplomés	Chefs	All
Percentage of friendships originating:	%	%	%	%	%
In youth	29	22	17	8	19
At professional school	13	13	33	50	28
At work	25	35	19	17	23
Through wife, kin	17	17	11	17	15
Through neighboring	0	13	3	0	4
Through voluntary associations	8	0	0	4	3
In other activities	8	0	17	4	7
Totals	100	100	100	100	99

	Techniciens	Ingénieurs autodidactes	TELECO Ingénieurs diplomés	Chefs	Cadres administratifs	All[a]	Both[b]
	%	%	%	%	%	%	%
In youth	27	6	16	18	45	17/22	18
At professional school	15	13	34	55	24	31/30	30
At work	19	31	16	9	10	17/16	20
Through wife, kin	23	13	12	5	7	13/12	14
Through neighboring	0	0	3	0	0	3/3	3
Through voluntary associations	0	0	3	5	0	2/2	3
In other activities	8	37	16	9	14	16/15	12
Totals	100	100	100	101	100	99/100	100

[a] The figure before the slash is the percentage for all categories combined, excluding the *cadres administratifs*: the figure after the slash includes them.
[b] These figures are the averages for the two companies combined, with the *cadres administratifs* excluded.

Table 9.3. *Friends' characteristics by rank and firm*

Friends	Techniciens	Ingénieurs autodidactes	Ingénieurs diplomés	Chefs	All
	PAMPCO				
	%	%	%	%	%
One or more work in same firm	50	86	33	22	44
Most or all in same occupation	25	50	58	67	50
None in same occupation	17	0	21	11	15
Most are *cadres*	8	88	78	89	64
At least one is a manual worker	8	25	11	0	10
Most or all live within 30 miles	73	86	53	45	61
	TELECO				
One or more work in same firm	23	33	32	11	27
Most of all in same occupation	23	56	58	67	52
None in same occupation	31	11	8	22	17
Most are *cadres*	15	43	78	100	63
At least one is a manual worker	0	0	0	0	0
Most or all live within 30 miles	62	45	58	50	54

autodidactes, most of whom did not meet their friends at a technical or vocational school, it could be regarded as additional evidence for some sort of occupational community.

Here, however, it is important to recall how heterogeneous is the work of engineers and how they tend to identify with each other more as fellow *cadres* than as fellow engineers. *Cadres* do not share a particular occupational situation and ethos so much as similar employment, juridical and social statuses. But as with certain occupations, this *cadre* axis of solidarity within the world of work spills over into life off the job. Indeed, the most striking figures in Table 9.3 concern the percentages of respondents saying that most or all of their friends are *cadres*.[1] At both firms roughly four-fifths or more of the *ingénieurs diplomés* and *chefs* say most or all of their friends are *cadres*, but the figures for the *techniciens* are dramatically lower, especially at PAMPCO (8 percent). In short, the *cadre* barrier seems to affect social relations away from work as well as on the job. Moreover, although this barrier appears slightly less pronounced in the advanced technology setting, it is strong enough there as well to constitute evidence against Mallet's expectations of a growing class solidarity between *ingénieurs* and *techniciens*. Finally, the figures on friendships with manual workers strengthen the case against an emerging "new working class." Not a single engineer in the high technology sample had even one blue-collar friend. At the traditional site, the only significant departure from this pattern is among the *ingénieurs autodidactes*. The explanation there is that many of these men are themselves former manual workers. At TELECO, by contrast, there are few manual workers, and the *ingénieurs autodidactes* tend to have started their careers as *techniciens*.

Why do *cadres* socialize almost exclusively with other *cadres*? One reason, no doubt, is simply that their incomes are considerably higher than those of manual and lower white-collar workers. Indeed, the differential is larger than in other advanced countries. Moreover, *cadres* know that their incomes are going to grow over time. Their patterns of consumption (and also savings and investments) reflect this shared reality, contributing in the process to the formation of what could be called a *cadre* life-style – appropriately pegged to the stage of the career – that reinforces the experience of membership in a stratum apart, a stratum distinguished by its standard of living and leisure activities as well as the position of its income earners in the system of production.

A second reason that the *cadre–technicien* barrier affects relations off the job is that there are no competing axes of identification. This absence is best understood by contrasting it to the situation in the United States. There, where powerful unions in basic industries have enabled some older manual workers to earn and save enough to buy homes in "middle-class" residential areas, manual workers often share neighborhoods with administrative workers, professionals, and small businessmen (Halle, 1984). Moreover, the decentralized character of the American polity and the salience of ethnic and racial divisions combine to unite blue- and white-collar residents in defense of their neighborhoods. In centralized France, by contrast, neither local school board politics nor ward politics serve to integrate residents of the same neighborhood. Moreover, few *cadres* are members of a struggling religious, ethnic, or other minority group (women, handicapped, gays, etc.) that might integrate them with non-*cadres* in common struggles against discrimination.

It would be easy to imagine a form of privatization developing under these conditions, and indeed, the non-work life of many of these engineers is largely family-centered. For example, judging from the data in Table 9.4, most of the engineers spend most of their leisure time alone or with their

Table 9.4. *Leisure activities by firm (multiple replies)*

Percent who mentioned:	PAMPCO	TELECO
None but relax with family	2	3
Gardening, home projects	44	33
Individual sports	8	10
Team sports	2	4
Social sports, games	9	16
Visiting, going out	15	19
Cultural activities	13	9
Voluntary associations	3	3
Other activities	4	4
Totals	100	101

immediate family. Much of the gardening and home-repair activity is viewed as mildly obligatory, but it is also often embraced as a welcome respite from the tensions of work life. Yet it would be a mistake to interpret such involvement in family and home as the retreat of beleaguered individuals into the isolated haven of the nuclear family. Rather these engineers positively value their families and homes as good in themselves. Moreover it is evident from the previous discussion and from Table 9.4 that many of them have close friends outside of the family and participate in a variety of social activities.

The last type of evidence that is useful here is that regarding participation in voluntary associations. It is clear from Table 9.5 that the majority of these men are actively involved in groups and organizations outside their families. Quite a few of them are officers of these associations. For most subsets of interviewees on most of the measures, the participation figures are higher at TELECO. This is especially the case for the percentages involved in company-sponsored associations. The most likely reasons for this are first, that the stronger employees' association there (effectively run by the left-wing unions) organized teams and clubs for a much greater variety of activities than offered at PAMPCO, and second, that many more of the PAMPCO engineers (65 percent vs. 18 percent) lived "in town" – a small company town much less – making informal interaction with workmates and others relatively easy. A separate analysis of these data that distinguished "locals" and "out-of-towners" showed that at both companies the out-of-towners were considerably more likely to belong to and be active in voluntary associations. Finally, it is noteworthy that many of the interviewees – 43 percent at TELECO – participate in a company-sponsored association, but that none of the PAMPCO *techniciens* do (in contrast to 55 percent of the *techniciens* at TELECO). This latter finding is one more sign of the hardness of the *cadre* barrier at PAMPCO, and the sense of alienation of a

Table 9.5. *Voluntary association participation (in percentages)*

| | PAMPCO | | | | | TELECO | | | | | |
	Tec	IA	ID	Ch	Average	Tec	IA	ID	Ch	Adm	Average
Belonging to at least one	67	50	74	50		69	73	64	90	67	
Belonging to more than one	8	10	21	20		31	45	36	70	27	
Active	58	44	58	56		62	73	60	82	53	
Officers	25	33	6	22	19	27	30	21	63	14	27
At least one company-sponsored	0	22	6	22	10	55	50	47	38	27	43

Tec = *Techniciens*
IA   = *Ingénieurs autodidactes*
ID   = *Ingénieurs diplomés*
Ch   = *Chefs de service*
Adm − *Cadres administratifs*

stratum that was excluded from the company's officer corps, yet disinclined to participate with other workers in a predominantly blue-collar setting.

To what sorts of associations did the engineers belong? Table 9.6 shows the percentage distributions by type of association, rank, and company. The most popular type of voluntary association among these men is the recreational type. Included here are soccer teams, bridge clubs, and photography associations, the latter entailing little social involvement. Nevertheless, significant minorities of engineers are involved in the more serious types of associations, types that could have political implications. In examining these distributions, it is helpful to keep in mind how much they reflect life-cycle factors. Thus, the older categories of engineers – the *ingénieurs autodidactes* and the *chefs de service* – are less likely to be involved in sports, but more likely to have children in school. With respect to participation in religious groups – usually Catholic lay associations – the figures in Table 9.6 reflect the well-established relationship in France between social class and church involvement. The higher level of participation in civic and political associations at PAMPCO probably reflects the facts that Pamtown is a small town in which many of the PAMPCO engineers lived as well as worked, and that PAMPCO's management paternalistically encouraged "responsible" participation in local affairs.

What conclusions may be drawn from these findings on the friendships, leisure activities, and voluntary associations of these engineers? First, it seems clear that the non-work lives of these men are not accurately characterized by such concepts as "privatization." Despite the fact that they value

Table 9.6. *Types of voluntary associations (percents)*

	PAMPCO					TELECO					
	Tec	IM	ID	Ch	Average	Tec	IM	ID	CH	Adm	Average
Child/PTA	0	17	6	29	11	0	38	4	11	0	10
Religious	0	0	19	14	11	0	0	4	17	7	6
Sports, games	44	33	63	29	47	85	54	76	39	64	64
Civic, political	44	50	0	29	24	8	0	12	17	21	12
Other	11	0	13	0	8	8	8	4	17	7	8
Totals	99	100	101	101	101	101	100	100	101	99	100

the familial side of their lives and seek to protect it from work-based intrusions, they do not limit their participation in the larger society to work alone. Through friendships, recreational activities, and voluntary associations, they are involved in a wider network of social relations.

Second, these relationships are rarely the kind that could form the basis for integration into residential communities. Integration into residential communities is inhibited by the weakness of local decision-making bodies, the lack of ethnic and racial segregation, the general tendency of the French to be wary of "others" (Wylie, 1964: ch. 9), and the fact that young careerists tend to change residence often, either because their careers require it or because increasing salaries permit moves to more expensive housing.

Third, many of these engineers' friendships and activities link them to persons employed in the same company, especially at PAMPCO. Yet, the patterns of integration are quite different at the two companies; while recreational activities sponsored by employee associations effectively promote friendly contacts between *ingénieurs* and technicians at TELECO, *cadre* housing and the predominance of informal contacts reinforce the exclusiveness of a *cadre* community at PAMPCO. Finally, there is little in the evidence to suggest the existence of an occupational community among these engineers. At PAMPCO, engineers associate with one another and with *cadres administratifs* as *cadres*, a multi-occupational stratum that subsumes engineering; at TELECO, the community of all technical workers is stronger, but still divided by the privileged associations of *ingénieurs diplomés* with each other and with graduates of the same school. The latter would seem to be of increased importance, especially for graduates of prestigious schools, for the expectations and probabilities of changing firms are higher, thus making "old school" networks more critical to careers, and the greater number of *petite école ingénieurs* stimulates status distinctions.

Finally, however, the data on friendships suggest the presence of a *cadre* stratum outside the work place. In this sense, work place divisions do spill over into the larger society, although not in the manner suggested by either neo-Marxian theories or class or Durkheimian theories of occupations. For one thing, *cadres* remain aloof from blue-collar workers, and even from *techniciens* at PAMPCO. More importantly, it is not the content of the work *cadres* do that distinguishes them so much as their employment status and market position. Whether *cadres* constitute a service or professional class, however, remains to be seen.

Political participation

In examining political beliefs and behavior, it is useful to distinguish between content and process. Content refers to the actual stands citizens take on partisan issues. The two sections following this one investigate the content of French engineers' voting behavior and political ideology. Process refers to the intensity of concern and to the extent and form of participation in the politics, and is the subject of this section.

There are two major ways by which employed citizens participate in politics in democratic states. One is through their unions (or professional associations), with the unions engaging in such political activities as endorsing candidates, lobbying for legislation, and publicly advocating certain governmental policies. In joining a union, a French worker is normally making more of a political statement than is his or her American counterpart. One reason is that union pluralism and the lack of "closed shops" in France make membership in a union a more voluntary act. Furthermore, French labor unions are more politicized than American ones; some have close ties with major political parties. The other way of participating is through direct participation in the political process, with the range of action extending from simple voting to participating in electoral campaigns or even running for political office. Since the previous chapter examined union activity, this section focuses on direct participation by individuals.

Each of the interviewees was asked: "Are you interested in politics?" (followed by a probe about voting habits), and then "Have you ever happened to do more than simply vote, for example attend political meetings, participate in an electoral campaign, sign a petition, write a letter, or participate in a political demonstration?" Only one respondent, a *technicien*, said he never voted; seven others admitted to voting only irregularly. Many expressed contempt for politics and politicians, but only two expressed the kind of outright cynicism evident in the following replies to the question about interest in politics.

Not really. I vote more *against* than for things.

No. No, I haven't voted lately. They only want your vote, and don't care. Besides, one can't really know a person's politics or what he'll do once in office . . . I feel I'm being had by the current voting system.

Significantly, both of these respondents were *techniciens*, one at each firm.

Most interviewees, however, said they were moderately interested in politics, followed the political news in the mass media, and discussed politics during electoral campaigns. At the same time, many displayed a disdain for politics, especially the "everyday politics of party squabbles and public theatrics." In the words of one, "I'm somewhat interested, but don't really like politics; it's too partisan, conflictual, and unconstructive." Another said: "I don't like politicians, but it is important to vote. They make decisions about laws and taxes." No doubt, the high socioeconomic status of *cadres* gives them a stake in political decision-making that fosters informed participation, but there appears to be a normative explanation as well for this ambivalent involvement. As one PAMPCO *chef* put it: "I willingly vote; it's a serious duty. I don't like the political games and lies, but I live in society, so am concerned with major issues and must be informed to vote intelligently." It may be that *cadres*, who often defend their privileges on the grounds of their responsibilities, feel a particular obligation to act the role of a responsible citizen.

Most of these engineers limited their political participation to simple voting, but a substantial minority of them – about 40 percent – had taken part in the political process in some more active manner. However, much of this participation is of a non-partisan kind. The two most common forms of participation mentioned were attendance at *réunions d'informations*, i.e. meetings where candidates debated or gave campaign speeches, and assistance with poll watching or vote counting. Most of the respondents emphasized that attendance at the meetings was simply "to find out about things," not to show support for a candidate.

To be sure, there were several engineers who were more active, but this activity too was often more non-partisan in character than evident at first glance. Consider the case of the seven engineers who served, or had run for seats, on their town council. In the United States, that level of involvement in local government would indicate a political activist. In France, however, town councils make few controversial decisions, and serve rather the symbolic function of providing democratic legitimation for decisions made by ministers, prefects, and subprefects.[2] Local mayors may acquire some influence, but "[i]n general the other members of the municipal council are quite content with an at least outwardly rather passive role of leisured notables who regard their election as reward for social status and economic success" (Ehrmann, 1983: 107). Council meetings are rarely attended by members of the public, and in smaller towns, the council elections are often nonpartisan, even uncontested. All this squares with the reports of the interviewees who were councilmen, all but two of whom emphasized the apolitical and civic nature of their roles.

There were, of course, some interviewees who had been or were engaged in partisan politics. Several had signed petitions; some had helped circulate them. About a half dozen said they had participated in political campaigns. Such participation reflected concerns based on a variety of identities: as *cadres*, as workers, as consumers, as parents, as Catholics, etc. The following comment is illustrative:

Yes, I've been moderately active in recent campaigns in order to prevent a victory of the left and thus the nationalization of *écoles libres* (private, usually Catholic schools) [a *chef* at TELECO's regional factory, a practicing Catholic, and the father of three school-age children]

In France, of course, such concern about education neither reflects nor reinforces membership in a residential community, for educational politics are a form of national, not local politics.

Among the interviewees who were activists, those supporting right-wing candidates were *ingénieurs autodidactes* and *chefs*; those supporting left-wing candidates and causes tended to be young *ingénieurs*, including *ingénieurs commerciales* (sales engineers), at TELECO's Simulator Division. Most of the latter were student militants before begining full-time employment. Indeed, many of the younger interviewees had marched in demonstrations as students in the late 1960s, but few had since then. One former activist working at PAMPCO admitted that fears for his career had been one reason he quit being "a militant." Yet, like most others, he also pointed to the time problems created by work and family responsibilities, and confessed a certain loss of faith in the potential for radical change through such activity. Even the engineers who were active hastened to stress that they were not militants. Only two belonged to a political party, and they were *ingénieurs autodidactes* affiliated with centrist and right-wing parties.

In concluding, a few points bear emphasis. First, the *cadres* at both firms in this sample display a fairly high rate of participation in the political process; even the *techniciens* reveal only limited tendencies towards cynicism and privatization. Second, the participation of those who do more than simply vote tends to be of a moderate and civic rather than militant or partisan kind. Third, the concerns that stimulate participation grow out of both work and non-work roles, out of their statuses as parents, home-owners, and even former students, as well as their positions as *cadres*. Yet, their career prospects appear to reinforce the inclination towards moderate and largely non-partisan political activity. In short, the picture that emerges here complements that emerging from the examination of social partici-pation: neither embattled nor withdrawn, French technical workers appear to be well integrated into contemporary society.

Table 9.7. *Percentage voting right by rank and firm*

Firm	Techniciens	Ingénieurs autodidactes	Ingénieurs diplomés	Chefs de service	Cadres administratifs	(N) Total
PAMPCO	30	100	86	90		76(32)
TELECO	36	38	43	90	70	53(32)
Totals	33(7)	69(11)	60(21)	90(18)	70(7)	63(64)

Party preference and voting

Although there were few differences between the PAMPCO and TELECO engineers in the levels and character of their political participation, there were major differences in the candidates and parties they chose to support. Moreover, these differences are of the kind that new working class theorists anticipated. Mallet maintained that the character of technical work in advanced industries was undermining the *cadre* barrier between *techniciens* and lower level *ingénieurs*, and bringing them together in common opposition to their superiors. Furthermore, he thought this opposition to capitalist domination would extend beyond the work place, leading to similarly left-wing politics by both groups.

Each interviewee was asked what political party he felt closest to, and how he had voted in three recent elections.[3] In view of the very high correlations among the responses to the three national voting questions, the analysis below makes use in places of a dichotomized scale that gives equal weight to each of these items.[4] Table 9.7 presents the percentages voting for the political right on this scale, by rank and firm.

Two features of this table stand out. First, much larger proportions of *cadres* than of *techniciens* support the right, especially at PAMPCO. Second, support for the right is much lower in the high technology setting, with the decline concentrated among the *ingénieurs autodidactes* and *ingénieurs diplomés*. This section explains and interprets these variations, especially the greater support for the right at PAMPCO.

It is well known that one of the best predictors of voting behavior is the voting practices of parents, and this is certainly true for the interviewees.[5] However, this seems to have little bearing on the political differences between PAMPCO and TELECO engineers. Similar percentages of the engineers at each firm – 76 percent at PAMPCO and 78 percent at TELECO – said that their parents favored parties of the center-right. Moreover, the differences among ranks were small and contrary to expectations. Only among the *ingénieurs autodidactes* and *chefs* at PAMPCO and the *techniciens* at TELECO did substantial proportions (36–43 percent) say their parents supported the left.

Table 9.8. *Party preferences by firm (in percentages)*

Firm	*Majorité*	Ambivalent	Left	Total
PAMPCO	55	29	16	100 (49)
TELECO	39	27	34	100 (56)
Total	47(49)	28(29)	26(27)	101(105)

In view of the fact that 59 percent of the *ingénieurs diplomés* at TELECO are graduates of *petites écoles*, as opposed to 35 percent at PAMPCO, it might be thought that the political differences can be explained by this factor. In fact though, a slightly higher proportion of the 20 *petite école ingénieurs* in this subsample – 65 percent as opposed to 58 percent for the graduates of *moyennes* and *grandes écoles* – vote for candidates of the right. Similarly, it might be imagined that the social origins of the engineers at TELECO are lower than for those at PAMPCO, and that this difference helps explain the political profiles. However, while it is true that father's occupational status is strongly associated with son's voting in this sample, the percentage of fathers who were *cadres*, professionals, or semi-professionals at TELECO was 58 percent, as opposed to 45 percent at PAMPCO. A greater proportion of the parents of TELECO engineers were also self-employed in small, family businesses, while only half as many – 11 percent vs. 23 at PAMPCO – were manual workers.

In trying to understand the politics of these men, it is helpful to examine the distributions on indicators that do not force a choice between left and right so clearly as the voting scale. Party preference is one such indicator, for it was coded into seven categories.[6] In view of the overall distribution on this variable – only one Communist, for example – these responses were regrouped into three categories: (1) *la majorité* (three categories, including Giscardiens and Gaullists), (2) ambivalent (one category), and (3) the left (three categories). Table 9.8 presents the distribution by firm. In this table the differences between firms are less pronounced than in Table 9.7, and there emerges a substantial group in the middle. Moreover, the responses to the voting in the first round of the 1978 parliamentary election indicate that more than two-thirds of those favoring the majority preferred Giscard's modernizing, center-right party (UDF) to the more right-wing Gaullists (RPR)). Nevertheless, the left-wing support remains considerably higher in the high technology firm.

How do these findings compare with those from national surveys? It is well established that *techniciens* vote for the left in much greater numbers than do *cadres*. For example, in the first round of the 1978 parliamentary elections, 21 percent of *techniciens* voted for Communist Party candidates,

66 percent for various candidates of the left, including communists. By comparison, 35 percent of *cadres* (including *ingénieurs*) cast their ballots for candidates of the left (Capdevielle et al., 1981: 308). *Ingénieurs* appear to be less conservative than *cadres* in general, but only slightly. Thus, the proportions of *cadres* and *ingénieurs* who declared themselves closer to the left in 1974 were 21 percent and 23 percent respectively. By contrast, the proportion of *techniciens* was 40 percent (Groux, 1983: 112). (Among *ingénieurs* whose fathers were manual workers, the figure rises to 36 percent; for *techniciens* it rises to 47 percent.)

For information on variations by industry, the best source is Grunberg and Mouriaux's (1979) national survey of 1,481 French *techniciens*, *ingénieurs*, and *cadres*. Their findings reveal patterns very similar to those found at PAMPCO and TELECO. In the electronics industry, the proportions favoring the left and right are about equal, but in metal-working, the proportion favoring the left is less than half that favoring the right. Grunberg and Mouriaux suggest a couple of possible explanations for this difference. One is the different distributions of functions in these two industries, combined with the pronounced tendency for engineers in research and design to be more favorable to the left than are their colleagues in production. For example, they found that a higher percentage (44 percent) of *ingénieurs* working in research voted for Mitterrand in 1974 than of *techniciens* in production (34 percent), in spite of the fact that *techniciens* in general were much more likely to vote for the left (1979: 123–132). Within research, the *ingénieurs* and *techniciens* were equally likely to have voted for Mitterrand, but within production a gap opened up between them, with the *ingénieurs* (26 percent) moving even farther to the right than the *techniciens*. Grunberg and Mouriaux go on to note that *cadres* have traditionally supported the right much more than has the electorate in general, but that during the 1970s there was a considerable shift to the left, primarily among lower level *cadres*.

These findings appear to support Mallet's new working class thesis. However, after observing that support for the left is also high in such traditional industries as construction and "other services," Grunberg and Mouriaux offer a different explanation. They suggest that in traditional, mass-production industries, but not construction and services, many *cadres* supervise manual workers, and that the experience of difficulties in this aspect of their work generates conflict with workers' unions and sympathy for the forces of order. They also point to self-selection and the higher levels of educational certification among research *techniciens* and *ingénieurs* as contributing factors. The higher levels of certification are viewed as making those who have them less dependent on their local bosses and thus freer to express unpopular ideas. Finally, they note that because of the recent rapid growth of the high technology sector, the *techniciens* and *ingénieurs* there tend to be younger, and that youth is associated with sympathy for the left. At the same time, they acknowledge that within the traditional industries, *cadres* under 35 years of age offer no more support to the left than do their elders.

We are tempted to conclude that the political homogeneity which exists among the top managers of these [traditional] industries, the small proportion of *cadres* among all employees, and especially of young and new *cadres*, the presence of the *patronal*, and perhaps the criteria of recruitment, all facilitate the reproduction of right-wing political views in these settings. (Grunberg and Moriaux, 1979: 131)

This interpretation is compatible with much of the analysis presented in this study. PAMPCO was certainly a company that employed relatively few *cadres*, but took great care in the selection of them, and reminded them often of their social roles as company officers. The research at PAMPCO and TELECO revealed differences in the authority situations of research and production engineers, and in the related attitudes towards working-class unions and parties. For engineers, at least, the principal consequence of the emergence of high technology industry may be a disproportionate growth in the number of them working in research and design positions that involve no supervision of manual workers. This development is likely to have a greater effect on the politics of engineers in France than in countries where engineers have traditionally served in staff rather than line positions (e.g. Great Britain), or faced a less hostile working-class movement (e.g. the U.S.). Moreover, in France, the growth in the proportion of *techniciens* and *ingénieurs autodidactes* who have intermediate level diplomas is undermining the traditional dependence of *cadres autodidactes* on their employers. The sample sizes are too small to permit further analysis of the *ingénieurs autodidactes* at TELECO, but it is noteworthy that the largest inter-firm difference in Table 9.7 is among the *autodidactes* (100 percent at PAMPCO voting for the right vs. 38 percent at TELECO), and that the TELECO *autodidactes* do possess more nationally recognized educational credentials.

The discussion of political participation through unions pointed to the different market situations of the metal-working and electronics industries, and the implications for engineers' attitudes towards various parties. In this context, it is noteworthy that the greatest support for leftist parties among French *cadres* is found among those working in the public sector, including the nationalized industries and firms. In the 1978 parliamentary elections, for example, 58 percent of public sector *cadres* (excluding teachers) voted for the left, but only 21 percent of private sector *cadres* joined them (Capdevielle et al., 1981: 308). It is in the labor market interests of public sector workers to support parties of the left, for those parties are committed to increasing rather than cutting government services and budgets. For similar reasons, the workers at TELECO had an interest in expanded government expenditures, especially of the kind promised by the Socialists. The major customer of TELECO was the French government; the P.T.T. bought its telephone systems, the military and nationalized airlines bought its flight simulators. Moreover, the Socialists were pledged to increase support for research and development in the high technology industries of the future.

Yet there is more to the matter of market situations. For the *ingénieurs diplomés* at PAMPCO, the external labor market was indeed discouraging; 50 percent of them, compared to 22 percent of those at TELECO, said it would be difficult to find a similar job elsewhere. At the same time, the small numbers of *ingénieurs* at such "blue-collar" companies – 2.7 percent at PAMPCO – means that their salaries account for a small portion of the overall wage bill. Thus, their employers can afford to pay them well, even in hard times, and at PAMPCO they did so. By contrast, at TELECO, where the *ingénieurs* constituted 10.2 percent of the work force, their salaries – especially those of the young *ingénieurs* – were considerably lower. Moreover, this was a major grievance; few complained of boring jobs or insufficient autonomy, but 63 percent said the engineers in their company were underpaid, compared to 37 percent at PAMPCO. In addition, almost three times as many of their wives worked in clerical or blue-collar jobs, despite the fact that these women were better educated in general than were the wives of the PAMPCO *ingénieurs*.

Finally, feelings of job insecurity were just as widespread at TELECO as at PAMPCO, despite the former's difficulties in hiring all the engineers it needed in some divisions, and the latter's recent operating losses. The reason seems to have been the project nature of production at TELECO, combined with the large numbers of engineers, the company's ongoing reorganization, and the related lack of company commitment to its *cadres*. At paternalistic PAMPCO, the company expected much of its *cadres*, including conduct becoming a *cadre* off the job and – implicitly at least – in politics, but in return, PAMPCO took care of its *cadres*. As in many old French companies, management there had never dismissed a *cadre* for economic reasons, and seemed determined not to. It surely was because of such commitment to the staff, as well as the higher salaries paid at PAMPCO, that 56 percent of the engineers there said the company does more for its employees than necessary, while none did so at TELECO. In short, the engineers at TELECO were unhappy about their pay and security, and blamed company management; the engineers at PAMPCO were unhappy about their careers – 48 percent are disappointed with them to date, as opposed to 28 percent at TELECO (and only 31 percent believe they have a good chance of accomplishing their career goals, as opposed to 57 percent at TELECO) – but they blamed the economy, or the government, or the workers' unions, not management. Such patterns of frustration and blame generate support for candidates of the right rather than the left.[7]

It is important to recognize that the lower pay of TELECO engineers was not a direct consequence of the larger number of technical positions at that high technology firm. It had to do also with how the company chose to fill those positions, and the availability of certain kinds of labor, due to autonomous developments in the educational system.

Filling them with *ingénieurs diplomés* has resulted in a substantial growth in the proportions of company personnel who are *cadres* in the industries that are hiring technical workers. It is hardly surprising that at the early stages of their careers, these *cadres* are paid less than the average for *cadres*, even young *cadres*, in all industries together. Nor is it surprising that in a country where *Monsieur l'ingénieur* was until recently both well paid and a commander of subordinates, some of these young *cadres* are unhappy. In time, their careers, like Pierre's, will carry them to middle management positions that pay well and involve supervision, and judging by the more conservative responses of the *chefs de service* in both firms, and of Pierre 10 years later, the young *ingénieurs'* politics will adjust accordingly.

Political ideology

Voting behavior is often a poor indicator of underlying political values. This is especially so for voters oriented towards the center of the political spectrum in a society polarized into right-wing and left-wing blocks. To understand the political ideology of such citizens, it is useful to examine more direct evidence of their attitudes. Thus, this section analyzes the responses of the interviewees to several conventional questions about public policy, including questions about taxes, public spending, economic growth, and the nationalization of private enterprises.

Each interviewee was asked (question 22) if any group in France had too much power. While 43 percent of the TELECO engineers said the rich, employers, or something comparable, only 17 percent of the PAMPCO engineers gave similar replies. PAMPCO engineers were more likely to point to the left-wing unions and the Communist Party (21 percent) or to central government (21 percent). Here then is additional evidence of more favorable attitudes towards the left among engineers in the high technology setting. Again, however, the pattern of replies fits well with the argument made above: that in France radical left-wing unions evoke anti-left sentiments among *cadres* in labor-intensive settings, that *cadres* in old industries fear that government intervention in the economy will favor blue-collar workers and the high technology industries of the future, and that selection and self-selection professes generate more conservative *cadres* in traditional, paternalistic firms. The pattern of replies also reveals differences among the various categories of engineers. Only 8 percent of the *techniciens* named the unions and parties of the left, compared to 40 percent of the *ingénieurs autodidactes*, and 44 percent of the *cadres administratifs.*

Each interviewee was also asked, in a follow-up questionnaire, about the areas in which government action had the greatest impact on his life.[8] The replies display few differences between the two firms or among the different categories of engineers. Yet, they are noteworthy for the widespread sensitivity they reveal to government's economic impact. In a list of sixteen areas, taxes were singled out by almost a quarter of the respondents at each firm, and inflation by almost another quarter. The third most often mentioned area was employment and career. Eleven percent of the PAMPCO engineers and 18 percent of those at TELECO mentioned this as one of the three areas in which the government had the greatest impact. Twenty-nine percent of the *ingénieurs diplomés* at TELECO gave this as the area in which government had the most impact. By contrast, very few respondents mentioned the government's role in maintaining democracy, protecting liberties, or promoting equality, and none mentioned protection against crime, preservation of the environment, or enhancement of the moral climate. It would be a mistake to overinterpret these data, but it is worth noting that in a country known for its ideological politics, these engineers tend to react to government in terms of its effects upon their individual economic well-being, in general and, especially in the case of the high technology firm, its impact upon their careers.

The concern about taxes is suggestive of the tax revolts in America. However, an interview question (number 25) reveals a rather different set of attitudes. The main complaint is not that the tax system is biased against the middle class, but rather against the poor. Only 16 percent of all respondents judged the French tax system as "anti-middle," while 46 percent viewed it as unfair to the poor, and 39 percent regarded it as fair. There were significant differences between the two firms, and among the different categories of engineers. Only 32 percent of the PAMPCO engineers but 57 percent of their TELECO counterparts viewed the tax system as unfair to the poor. Half of the *chefs de service* and more than half of the *ingénieurs autodidactes* regarded the current system as fair, as did 38 percent of the *ingénieurs diplomés.* By contrast, only 18 percent of the *techniciens* agreed. Here again is evidence of reduced defensiveness about traditional *cadre* privileges in the high technology setting, but also of a significant difference in attitudes between *techniciens* and *ingénieurs.*

It is not only with respect to taxation that the *techniciens* give evidence of belonging to a different social stratum than *ingénieurs*, with correspondingly different conceptions of their interests. Similar differences emerge with regard to public spending. While there were no differences between the two firms, many more *techniciens* (52 percent) favored cutting defense spending than did *ingénieurs diplomés* (33 percent), *chefs de service* (12 percent), or *ingénieurs autodidactes* (10 percent). What the *chefs* (65 percent) and *ingénieurs autodidactes* (50 percent) wanted cut was waste in government operations. These large differences are unrelated to social origins or parents' politics.

Another subject on which the *techniciens* stood out was that of economic growth. While 59 percent of *techniciens* favored slower economic growth, only 29 percent of *ingénieurs diplomés*, and 17 percent of *ingénieurs auto-didactes* agreed; none of the *chefs* agreed. In explaining their attitudes, over half of the *chefs* emphasized the economic threat of foreign competition and the importance of France not falling behind in the international race.

The most controversial political issue in France in the late 1970s was the nationalization of more large industrial enterprises. A major reason for the collapse of the Socialist–Communist alliance in 1978 was the disagreement between these two parties of the left on how many companies would be nationalized if they came to power. Among the interviewees, only 30 per-cent favored the nationalization of any additional companies. Twenty-eight percent thought that the existing degree of nationalization was satisfactory, but 42 percent felt that too many French firms were already nationalized. There were no significant differences between the two firms in these dis-tributions, but there were major differences between the *cadres* and the *tech-niciens*. While 71 percent of the *techniciens* favored additional nationaliz-ations, fewer than 25 percent of each of the other categories of engineers agreed. As with the issues of *autogestion*, the most common criticism of the further nationalizations was that they would reduce the efficiency and competitiveness of French industry.

Taken together, all this material suggests that French engineers value economic growth and view private enterprise as conducive to it. In these senses, they are quite committed to the existing economic system, much more so than the technicians. To be sure, there are differences between PAMPCO and TELECO, but they are smaller than suggested by the initial voting profiles, and appear related to the different situational problems especially market-related ones, confronting the efforts of the engineers at each firm to realize their common values.

Social class

In attempting to define the class position of such knowledge workers as engineers, sociologists have offered a bewildering array of concepts. These include a "new middle class," a "new working class," a "working middle class," a "new class," a "new petty bourgeoisie," a "professional class," a "professional-managerial class," and a "service class." It is beyond the scope of this book to explore the strengths and weaknesses of each of these con-

Table 9.9. *Self location in class structure (in percentages)*

Class	Techniciens	Ingénieurs autodidactes	Ingénieurs diplomés	Chefs de service	Cadres administratifs
Highest	8	0	11	29	31
Middle of 3	29	50	59	62	54
Third of 4, of lowest of 2 or 3	54	15	25	0	15
Other	8	35	5	9	0
Totals (N)	99(24)	100(20)	100(44)	100(21)	100(13)

cepts for grasping and making intelligible the ideology of French engineers. Thus, the study has focused on those emphasizing changes resulting from the rise of high technology industry. Having already discussed in some length the concept of a new working class, this section focuses on Goldthorpe's concept of a service class, and Bell's concept of a professional class. At the same time, it refers occasionally to the other concepts. First, however, it examines the engineers' own images of the class structure.[9]

Sixty percent of the interviewees said there are three classes in France. The majority of these viewed the class structure as shaped like a pyramid rather than like a diamond – the "middle-mass" model (Wilensky, 1960) – or some other form. Only among the *techniciens* did a substantial minority – 29 percent as opposed to 12 percent or less among all other categories – identify only two classes. In every category except the *techniciens*, some respondents said there were no classes.

Each interviewee was also asked to indicate the class in which he would place himself. Again, the *techniciens* stand out for the feelings they display of belonging to a lower class than the other engineers. Table 9.9 summarizes the responses. While 54 percent of the *techniciens* see themselves as in either the lowest class or the third of four classes, only 16 percent of the *cadres* agree. Moreover, only 2 percent of *cadres* see themselves as in the lowest of three classes, as opposed to 21 percent of *techniciens*.

One other question asked: "In your opinion, within French society, is a *cadre* (or *technicien*) like yourself closer to manual workers, office workers (*employés*), top managers (*dirigeants*), civil servants (*fonctionnaires*), or free professionals?" After eliminating those who said civil servants, none, or all, there remained ninety-three responses. Among those, the percentages saying they felt closer to manual or office workers were 24 for *chefs de service*, 31 for *cadres administratifs*, 52 for *ingénieurs diplomés*, 67 for *ingénieurs autodidactes* and 95 for *techniciens supérieurs*. Yet, there were also significant differences between the firms. While 48 percent of the respondents at PAMPCO felt closer to the lower categories, 70 percent at TELECO did.

This difference between the two firms is accounted for largely by differences in the proportions of *ingénieurs diplomés* who said they felt closer to office workers (none at either firm said manual workers). The figure is only 27 percent at PAMPCO but rises to 81 percent at TELECO. Here again is evidence of a decline in identification with management among *ingénieurs* in the high technology setting. This is not a matter of age or *petite* vs. elite *école*, but it probably does reflect the fact that there is a strong relationship between the stratum with which an *ingénieur* identifies and the number of subordinates he has. In any case, by the time the TELECO *ingénieurs diplomés* are *chefs de service*, they seem just as likely – indeed, more likely – to identify with top management as are their counterparts at PAMPCO.

In short, many French *ingénieurs* do not feel themselves to be part of management, their *cadre* status notwithstanding, until they occupy a managerial position. In the past, most *ingénieurs diplomés* occupied such positions from the beginning of their careers, and this remains the case in traditional industrial settings. In the science-based sector, however, there are so many engineers and so few manual workers that most *ingénieurs diplomés* must work the first few years of their career in staff positions that involve subordinates only in the loosest sense of the term. Neither they nor their colleagues are embattled with workers' unions on a daily basis. It seems that where there are many *ingénieurs* and few or no visible blue-collar workers to make them feel relatively superior, they begin to *feel* like workers themselves. That they must wait for promotion to a higher position, and that they work closely with *techniciens* and former workers (*ingénieurs auto-didactes*) in the meantime, appears to affect their politics.

Much of the material presented in this and preceding chapters appears to support a view of the *cadre* boundary as a class boundary between a service class of trusted workers and an intermediate stratum of technicians and similar employees. It is not simply that there are striking differences in the class images, political attitudes, and electoral behavior of *ingénieurs* and *techniciens*. It is also that the content of the *ingénieurs'* attitudes and the character of their political participation square with the expectations of Goldthorpe. They accept the existing economic system, but only because they view it as efficient. They emphasize the importance of economic growth, and insist on the *cadres'* need for collective organization to represent their specific group interests. Finally, the active but non-partisan character of their political participation suits their roles as trusted and responsible workers.

Yet, Goldthorpe's concept of the service class presupposes an excessively deterministic notion of those workers who *must* be trusted "when authority must be delegated" or "expertise must be drawn upon" (1982: 168). In other words, inherent in certain production processes are jobs whose occupants must be trusted because they cannot be controlled effectively. The problem here is that no simple logic of industrial efficiency defines a similar distribution of more and less trust-entailing jobs in industrial society. From management's perspective, it is desirable to have all employees act in ways that are consistent with organizational goals and values, and management will often redefine a given job as more or less autonomous, depending on the presumed trustworthiness of the occupant.[10] But different societies produce different percentages of persons who can, and do invest in a recognized reputation for trustworthiness, and who are thus socially defined as trustworthy individuals. Moreover, the same society may rapidly shift from producing a small percentage of such trustworthy workers to a considerably larger percentage.

Furthermore, there are important divisions among trusted workers that raise problems with any class analysis of their politics. The attitudinal summaries reported above show the *ingénieurs autodidactes* as consistently more conservative than the *ingénieurs diplomés*. Why? Goldthorpe acknowledges the current weakness of service-class consciousness and solidarity, but treats it as a passing phenomenon due to the recent expansion of the class and the resulting need to recruit members from lower social origins. By contrast, this study emphasizes the qualitatively different career positions of *autodidactes*, and the institutionalized dualism of the French system for recruiting trusted workers. The complementary relationship of its two parts, combined with the interests of employers in avoiding professional control of engineering, suggest a potential for persistence in one form or another. In the meantime, the *autodidactes*, lacking formal credentials, are entirely dependent on the good will of their employers, and tend to conform to the latter's conservative politics.

In addition to the different market situations and attitudes of *ingénieurs diplomés* and *ingénieurs autodidactes*, there is a growing stratification within the ranks of the *ingénieurs diplomés* themselves. The career opportunities for the graduates of *petites écoles* and elite *écoles* are almost as different as those of the *techniciens supérieurs diplomés* and *petite école ingénieurs*. Most of the *centraliens* will attain posts in top management – *la direction*; the *petite école ingénieurs* will not. It should be noted that in France, *la direction* is increasingly viewed by most *cadres* as a separate, powerful stratum between them and the faceless owners of capital. Similarly, below the *cadre* category, the *techniciens* are increasingly well educated and trusted with authority and responsibility. At TELECO, moreover, the young *ingénieurs* seem less trusted than at PAMPCO, the entire employment relationship between them and the employer being less long-term and morally significant. In these ways and in the attitudes expressed, the differences between the *cadres* and *techniciens* at TELECO are smaller than at PAMPCO, suggesting a declining differentiation.

Indeed, by some definitions of *cadres*, the *techniciens supérieurs* are counted among them. Two of the *cadre* unions – the CGC and the UGICT – welcome a wide variety of technicians, supervisors, foremen, and sales representatives. For years, the French census classified *techniciens* as *cadres moyens*, *ingénieurs* as *cadres supérieurs*, contrary to the practice within industry.

This uncertainty concerning the *cadre* boundary becomes more intelligible if the category itself is understood as a political construction rather than as a natural group rooted in the division of labor. Boltanski (1982) makes this point well in his account of the formation of the group, *cadres*. In this view, *cadres* are the social product of a collective project undertaken by *ingénieurs* in the 1930s in the face of threatening mobilizations by employers and workers. In order for engineers and middle managers to have a voice of their own in the increasingly centralized and state-sponsored bargaining forums, they needed state recognition of a formal organization that could claim to represent a substantial and distinctive category of employees. Thus, they mobilized all the "sounder elements" within industry. Once recognized, the category took on a life of its own, and became a social group. This category of workers is defined less by the nature of its work and careers than is the service class, and more by its members' political outlooks, though clearly the two are related. Its strength and character stand in marked contrast to "the weakness of a politically organized 'middle class' voice, supporting a programme which claims independence from capital and labour" in British industry (Smith, 1987: 73).

I would make just two criticisms of Boltanski's analysis of the *cadre* class.[11] First, in trying to locate *cadres*, it relies too heavily on Bourdieu's oversimplified scheme of those who dominate and those who are dominated. *Cadres* are not analyzed in terms of their own work, but rather in terms of the struggle between capital and labor; they organize to assure their position as junior collaborators within the dominant group, at the same time serving capital's interest by coopting many middle level employees who might otherwise join with the rest of organized labor.[12] Second, Boltanski's analysis is too macrosociological and too cultural. As such, it overlooks the changing place of *cadres* within the firm, and the significance of the firm's policies. And focused as it is on "cultural capital" and symbolic interactions, it ignores the content of the work *cadres* do. The heterogeneity that appears within the *cadre* category, even among those with similar levels of cultural capital, once one examines work and careers, casts doubt about any treatment of *cadres* as a class and a social group.

It was in the nature of such a loosely defined and high status group, however, to undergo a process of degradation over the years. Employers determine who is a *cadre* and therefore eligible for various benefits, and employers have found awarding *cadre* status to employees a cheap way to motivate and reward them. Employees cooperate readily in this inflationary process. Thus the group has been growing more rapidly than the kinds of posts it allegedly and originally represented. This satisfies the borderline individuals who reap its material and symbolic benefits, and increases the political weight of the entire group. Yet, as the group expands downward in a climate that is no longer conducive to "third force" solidarity, a climate in which the struggle between labor and capital no longer dominates politics, the category itself begins to underto internal stratification and differentiation. At the same time, the shifting boundary dramatizes the lack of correspondence between the division of labor and class boundaries.

There is yet another way in which *cadres* are divided, and it is here that Bell's conceptualization of a professional class is useful. According to Bell, the processional class in post-industrial society is fragmented in a number of ways, the most interesting of which is by "*situs*". By *situs*, he means the kind of organization – economic enterprises, government agencies, universities, etc. – in which people work. The argument is that the growing role of knowledge is altering the relative power of business, government, and science in such a way that the politics of class are yielding to a politics of *situs*.

In France, where *cadres* tend to stay with the same firm throughout their careers (Granick, 1972), and where the state actively intervenes in the economy (Hall, 1986), the political significance of industrial location is magnified. This is evident in the recent struggles within the main *cadre* union, the Confédération général des cadres (CGC), described in chapter 8. There, *cadres* in the depressed metal-working industries opposed the confederation's conciliatory posture towards the government of Giscard, and successfully fought for a policy of union support for the more protectionist Gaullists. These intense struggles between the "conservatives," "corporatists," and "modernists" reflected the different interests of *cadres* in different economic sectors. It is noteworthy that in France at this time, it was the conservative Gaullists that were associated with protectionist policies, contrary to the positions of Republicans and Democrats in the United States. All this suggests that industry matters, but not in the technologically deterministic ways often claimed. Rather, the significance of industry for an engineer's politics is contingent on the society's organization of technical careers and the national political context.

These divisions among *ingénieurs* and *cadres* call into question the utility of either "class" or "occupation" for capturing their social location and values. In looking to position in the division of labor, traditional versions of both concepts overlook the significance of career prospects for engineers' definitions of their interests and identities. To be sure, the Weberian concept of class emphasizes shared market capacity, but it assumes an unsegmented labor market in which workers of similar skill levels enjoy similar life chances. Similarly, class structure typologies that derive market capacity and class interests from possession of skills (Giddens, 1973; Wright, 1985) assume a free market in capitalism that provides returns to "productive assets" (labor, skill, etc.) commensurate with their productivity. Such approaches obscure the degree to which labor markets are socially structured and occupations socially constructed. They ignore the autonomy of the firm within capitalism, and the consequences for internal labor markets. They obscure the degree to which collective action in the political arena influences income and opportunity, both through effects on market shelters and through taxes and transfer payment policies.

In short, this study tries to show the utility of a "market situation" analysis that rather than presupposing an autonomous economy following its own logic, with social processes limited to the allocation of individuals in it, is sensitive to the ongoing dialectical relationship between historically formed groups and socially shaped economic processes.[13] Thus, I have analyzed the impact of high technology industry on the work, careers, and politics of French technical workers in the context of the society's specific *grande école–cadre* system for producing and employing engineers, and in the context of the work and labor market situation implications of the political and industrial relations systems.

The distinctively national character of technical labor markets and engineering interests is well illustrated by a brief comparison of class and politics among French and American engineers. Zussman (1985) convincingly interprets the politics of the American engineers he studied as those of home-owning, tax-paying citizens who share residential communities with independent professionals, public sector workers, and the upper levels of the industrial working class.[14] The result is that while engineers and workers view themselves as on different sides within the work place, they see themselves in the political arena as members of the same "working middle class," opposed on one side by the nonworking, tax-consuming poor, on the other side by the nonworking, tax-advantaged rich. Thus, neither position in the division of labor nor even employment relationship has much effect on life away from work, including politics. There are no basic social cleavages between trusted and other workers, nor significant differences in the politics of old industry and science sector engineers. The reasons are specific to the United States, including the separation of work and community, the decentralized character of American government, and the unionism of American workers.

In France, the situation is quite different. The *cadre* boundary at work shapes social relations off the job, but the character of recruitment to engineering, state economic intervention, party policies, and the industrial relations system generates a politics of industrial interests that divides engineers and inhibits both class and occupational solidarities.

10

Conclusions: technology, nations, and career structures

Heirs to the tradition of prediction – to Comte, Tocqueville, Durkheim, Marx, Tonnies, and Weber – sociologists expend considerably energy scanning the horizon for signs of emerging social trends and conflicts. Such searching often lends itself to overinterpretation of the significance of changes in the modern industrial order. This appears to be the case for assessments of the rise of science-based industries examined in this book. This concluding chapter selectively reviews the findings on the work and ideology of French engineers reported in the previous chapters, and discusses the theoretical implications. In the process, it elaborates this book's alternative approach to understanding recent developments among technical workers.

One of the striking facts about French engineers is that those employed in high technology industries display far greater support for left-wing unions and parties than do those employed in more traditional industrial settings. Studies of engineers in the United States (Zussman, 1985) and Britain (Whalley, 1986) reveal no such differences in those countries. The following two sections consider two kinds of explanations of such inter-firm differences in France.

Technology and the division of labor

Despite their differences, theories of proletarianization, professionalization, and a new working class all explain the values of technical workers by reference to the organization of production and the influence of knowledge technology on it. The findings of this study, however, indicate several problems with such arguments. Chapter 5 pointed to both similarities and differences in the organization of technical work at PAMPCO and TELECO. The major issues, however, are whether these differences in the organization of work reflect differences between low and high knowledge technology, and what the consequences are for the ideologies and solidarities of the engineers involved.

The main effect of the rise of high technology industries on the work of engineers is that a much higher proportion of engineers work in product development. Such work differs from traditional production engineering in that it involves long-term projects rather than daily rounds, and entails little or no supervision of manual workers but considerable collaboration with other development engineers. It is more similar to traditional methods and manufacturing process engineering in that it involves quite specialized forms of design work, but differs from them in the focus on new products rather than cheaper processes of producing standardized ones.

The significance of product development work is twofold. First, it involves the engineer in a much less conflictual relationship with production workers, for he neither supervises them nor deskills their work. Indeed, product development engineers have little contact with manual workers, but are themselves in some sense the direct producers of the value added to the raw materials used. Yet, the effects of reduced contact with blue-collar workers, and of becoming direct producers themselves, depend on the national context and the firm's policies. In France, where engineering is traditionally associated with supervising workers and where the workers' unions have long denied the legitimacy of capitalist authority, the effect of reduced contact is a decline in the hostility of engineers towards the workers' unions and demands. Furthermore, in France, where a bureaucratic and engineering mentality has long dominated top management's approach to organizing production, and where paternalistic employers have jealously guarded their managerial prerogatives in the face of a rebellious but weakly organized labor force, the effect of engineers becoming direct producers is a growing tension – at least initially – between management and engineers, a tension exacerbated by the sense of diminished status among engineers brought up on images of *Monsieur l'ingénieur*.

The second way in which product development is significant, however, neutralizes some of these subjectively proletarianizing effects. The high technology industries that are oriented toward product innovation and customization involve technologies and markets that are constantly evolving, thus giving rise to a continuing flow of new engineering problems and projects. Chapters 5 and 6 showed how the unpredictable and nonrecurring character of such problems and projects make deskilling the work of engineers quite difficult. In those cases where engineering tasks are effectively routinized, as sometimes happens in programming, testing, and maintenance, they are normally reassigned to "technicians," freeing "engineers" to use their special skills where still needed. This partly reflects the French tradition of defining engineers, but not technicians, as *cadres*, i.e. technical managers.

It is not only the difficulty of rationalizing high level technical work that protects engineers. Equally important is that employers may have no interest in deskilling such work, either in order to cut labor costs or to exercise tighter control over the labor process. Contrary to the assumption in proletarianization theory, capitalist competition is not limited to manufacturing standard products more cheaply than other firms. Especially in industries in which product development is important, companies may choose to compete by means of a highly flexible work force that can adjust rapidly to changes in technology or the market. Rapid adjustment usually requires high level technical workers who have the intelligence, training, and confidence to master new problems quickly. Flexibility may be achieved in one of two ways. A company can hire and fire specialists as needed, a strategy that American firms have often practiced. Or it may employ and keep engineers who are themselves adaptable because they are multi-skilled. The latter strategy may mean a higher wage bill than might be necessary otherwise, but yield lower hiring costs and engineers who are better known and more experienced and committed. This has been the strategy typically adopted in France, where most schools produce *polyvalent ingénieurs* (even the recent cohorts of *techniciens* are increasingly *polyvalent*) and where engineers tend to remain with one firm for their whole career.

Similarly, controlling the work of such experts is not only impractical, it is unnecessary, for engineers can normally be trusted not to abuse the discretion they need to do their jobs efficiently. Of course, management must find and maintain trustworthy engineers. This it successfully does in France by recruiting from the *grandes écoles*, selecting individuals with great care, classifying engineers as part of management (*cadres*) rather than the regular work force, and promising significant promotions to those who perform well.

TELECO's Simulator Division departed from these French practices by rapidly hiring many *ingénieurs diplomés*, who in many cases were graduates of relatively specialized *petites écoles*, and by offering them below average salaries and career prospects. Unsure that it could completely trust such engineers in jobs that affected the results of enormous team projects, and having large numbers of them to manage in direct production roles, management initially tended to treat these engineers more like technicians than *cadres*. As Patrick O'Hara (1985) notes, it took some time for the French telephone industry to learn how to manage and treat large numbers of engineers, workers whose expectations are for considerably more autonomy, authority, involvement in decision-making, and concern for their individuality, than expected by lower level employees. TELECO took a while to learn, and generated some hostility among its young engineers in the meantime.

Although not easily routinized, product engineering does involve the engineer more directly in the firm's relationship with its customers, for whom the products are often custom designed. This may limit the autonomy of the individual engineer but also provides a basis for resisting arbitrary demands by superiors. Moreover, the constraints experienced in such lateral coordination of work by project managers and customer representatives are visibly related to market demands, thus legitimating the authority involved.

The rise of high technology industries involves other shifts in the distribution of engineering specialties, such as growth in the proportion who are sales engineers. Yet, it is important to keep in mind that these various specialties have long existed, and that the impact of high technology production is on the distribution among them rather than the fundamental character of engineering work – more theoretical or routinized. Moreover, there remain considerable variations, even within the high technology sector. Thus, continuous-process industries such as petro-chemicals and nuclear energy employ larger proportions of production and maintenance engineers than does the electronics industry. Even within the latter, there are major differences between firms (or divisions) engaged in the large batch production of watches or calculators and firms (or divisions) engaged in the small batch or "one-off" production of highly engineered products designed to meet the particular specifications of industrial customers.

At the same time, the expansion of the electronics and telecommunications industries and the employment trends within engineering point to a second unfounded assumption that has plagued much thinking about modern industry. It is that industry is increasingly mechanized and capital-intensive, with the result that the role of human beings in production is reduced to operating and maintaining machinery designed by a few engineers. In Marxist terms, "dead labor" is progressively dominating "live labor", with major implications for class conflict and social evolution. In fact, production in many of the most advanced industries is *increasingly* knowledge- rather than capital-intensive. According to O'Hara (1985: 21), since the early 1970s the electronics industry has spent more on research and development than on material investments, and spends enormous sums – 10 percent of the entire wage bill at IBM – on training and retraining its workers. At the same time, the value added through the labor of traditional production workers has fallen to below 10 percent in several branches of electronics and telecommunications, and the proportion of workers in the electronics industries constituted by *ingénieurs*, *cadres*, and *techniciens* has continued to grow: from 25 percent in 1976 to about 33 percent in 1982.

This growing mass of highly skilled technical workers tends to evoke opposite and equally misleading reactions from engineers themselves and from observers of industry. Engineers begin to feel as though they are no longer a technical elite within a work force of largely unskilled and semi-skilled workers, but rather the workers themselves. Within their industrial sector, they have a point. Non-Marxist sociologists, by contrast, may interpret this disproportionate expansion of the technical staff as evidence of skill upgrading and the withering away of the stratum of alienated, unskilled worker in modern society. Both reactions ignore the disproportionate growth of the service sector, and large number of unskilled occupations within it. In terms of absolute numbers, the five fastest growing occupations in the United States include those of janitor, truck driver, and waiter–waitress (Pavalko, 1988: 333). If it ever made sense to regard the factory as society-writ-small, it certainly does not any longer.

To question the impact of high technology production on the division of labor and the significance of the latter, especially in any universal sense, is not, however, to view the rise of high technology industry as unimportant for the place and role of engineers. Rather, it is to suggest greater attention to market conditions, which are themselves influenced by trends in production technology.

Labor and product markets

This study has emphasized the significance of the market situations of engineers. The findings are entirely compatible with Larson's (1977) argument that the subordination of the market for engineering skills to markets for industrial products produced by economically powerful firms makes professionalization difficult. They are equally compatible with the idea that organizational careers attach individuals to their firms and larger societies, undermining the potential for collective protest by those occupying similar positions in the division of labor.

However, the focus of this study has been on the effects of variations in the market situations of French engineers on their industrial and political attitudes. The preceding chapters focused on variations by educational category and by industry. With regard to the former, the analysis revealed broad differences in the attitudes of *techniciens supérieurs*, *ingénieurs autodidactes*, and *ingénieurs diplomés*. Although the samples were too small to permit statistical forms of causal analysis, the interview data and the logic of the situations strongly suggested causal connections between the market positions of these categories and their differing views on a variety of topics.

Recall that 31 percent of *ingénieurs diplomés* felt their company did no more for its employees than the minimum required by law, while only 5 percent of *ingénieurs autodidactes*, but 61 percent of *techniciens supérieurs* agreed. Sixty-five percent of *techniciens supérieurs* felt that the company pay differentials were too large, compared to 48 percent of *ingénieurs diplomés* and 20 percent of *ingénieurs autodidactes*. Attitudes regarding the need for collective organization and towards particular unions varied widely among these different categories of engineers. The same was true for voting patterns, and attitudes toward public spending, economic growth, and nationalizations. The anger and radicalism of the *techniciens* seems clearly linked to their blocked mobility. The conservatism and CGC solidarity of the *ingénieurs autodidactes* surely reflects a combination of selection for loyal, hard work and dependence on their current employer.

This study has also emphasized the differences in the market situations of the engineers at PAMPCO and TELECO, differences related to both the product markets and the personnel policies of their companies. The PAMPCO engineers were carefully selected for desired character traits, hired under the assumption by both parties that the employment relationship was to endure for many years, and paid reasonably well. Promotion to managerial posts was quite likely, given the relatively small number of young *cadres* competing for them, but depended more on social relations and loyal service than at TELECO. Yet, these engineers worked in a declining and labor-intensive industry facing serious competition from foreign producers, even in the home market. In view of the sector-specific character of French engineering and the pronounced tendency of French firms to fill all but their lowest engineering positions through internal promotion, many of these engineers felt stuck and worried. Their politics, especially within the CGC, reflected this attachment to an industry whose only hope appeared to be in government policies of the kind promised by the conservatives, policies of protection for the industry rather than increases in the minimum wages of workers.

At TELECO, the situation was dramatically different. Recruitment was much less selective than at PAMPCO, salaries were lower, and the external labor market stronger. In such conditions, it is not surprising that the engineers felt less attached to the company, blamed top management rather than the government or working-class unions for their grievances, and dared act on them from time to time. The same reasoning would incline them towards greater support for the egalitarian and anti-management policies of the "non-categorical" unions and left-wing political parties. Moreover, the commitment of the French Socialist Party to major subsidies for France's high technology industries of the future naturally appealed to the engineers in this sector.

Such developments in the salaries and sentiments of engineers in the rapidly expanding high technology industries have led many observers to infer a proletarianization of engineers growing out of a massification and deskilling of technical work. This study has argued, however, that the disproportionate expansion of the *petites écoles d'ingénieur*, combined with the

French tradition of hiring all *ingénieurs diplomés* as *cadres*, and TELECO's decision to hire young *ingénieurs* instead of *techniciens* resulted in the emergence of a new, lower stratum in an expanded work force of engineers. As long as a constant proportion of the national labor force is engaged in highly skilled engineering, it is a mistake to attribute the emergence of new lower skilled "engineering" jobs to deskilling.

The effects of market situations are evident not only in strong associations with various attitudes, but also in the content of engineers' attitudes and in the lack of solidarity among engineers. Many engineers stressed their concern about advancement, even when trying to present themselves as above careerism. Some *ingénieurs diplomés* at PAMPCO preferred boring (in their minds) production jobs to interesting design ones for the avowed reason that the production jobs were better stepping stones to higher management. At the level of group formation, solidarity among French engineers – even among *ingénieurs diplomés* – is undermined by the fact that engineering is not a terminal occupation for many (and an unknown many) engineers, and that different educational credentials strongly affect life chances, although not so much as to create "communities of fate."

To emphasize career structures rather than the organization of technical work is not to deny the significance of changes in knowledge technology for the values of technical workers. It is rather to argue that a major and neglected influence of the rise of science-based industry on engineers is by way of its impact on their careers. In high technology settings, the high ratio of engineers to managers may mean fewer opportunities than in the past for engineers to move up into higher management. The rapid rate of technological advances means instability in product and labor markets, especially at the level of the firm.

Even here, however, generalizations about the impact of science-based industry must be made with caution, for it is difficult to disentangle the effects of technology from the effects of newness. High technology industries are new industries, and new industries, whatever their technology, tend to face different kinds of problems and opportunities than old industries (Stinchcombe, 1965). Because new industries offer unusual opportunities for capturing emerging markets, but also high risks stemming from intense competition and uncertainties about demand and costs, they tend to be unstable. Finally, even generalizations about specific industries often obscure the enormous variations among and even within firms in the same industry. The preceding chapters have emphasized the significance of several organizationally specific factors that seem to condition the impact of jobs on attitudes: organizational growth, bureaucratic structure, and management style, especially with regard to the recruitment, supervision, and promotion of technical workers.

Perhaps the only generalization that can be made about the rise of science-based industries is that they require many more technical workers than other industries, and that they employ larger proportions of them in technical specialties outside of the traditional, labor-intensive functions of production and maintenance. In any case, the social consequences depend on such factors as markets, career structures, and the preexisting social definition of an engineer. The demand for particular types of technical workers is influenced by managerial strategies concerning the recruitment, use, promotion, and dismissal of trusted experts, policies that vary within and especially among societies. Supply is shaped by demand, but also by several other factors, most notably the national educational system for producing technical workers, itself the outcome of struggles among engineers, employees and the state in the context of specific historical pressures and opportunities. It is these distinctively national factors that become important once the explanatory emphasis in accounts of engineers' ideology shifts from work to market situation.

The formation of a French structure of technical careers

In advancing explanations for variations in the positions and values of French engineers, the preceding two sections risk obscuring important similarities and the distinctively national organization of technical work and careers in France. In order to redress the balance, this section reviews the commonalities that characterize the employment situations of French engineers, and briefly examines their sources in French history.

The engineers at PAMPCO and TELECO share a common status as *ingénieurs* and *cadres*. Whether *autodidactes* or *diplomés*, they enjoy the traditional high status accorded French engineers, a status rooted in the well-established association between the occupation in France and the prestigious state engineering schools. As *cadres*, they enjoy a legal employment status that entitles them to separate representatives in work place elections, to participation in a special supplementary retirement fund, and to additional privileges specified in industry-wide employer–union agreements. As *cadres* and *ingénieurs*, they do highly skilled technical work within large industrial enterprises, enjoy autonomy on the job, and have careers that lead to progressively higher levels of pay and responsibility, especially if the enterprise is competitive and growing.

These shared statuses generate certain interests and attachments. The findings presented in the preceding chapters show that the engineers at both firms value efficiency and economic growth, view the market as a natural mechanism for achieving both, and thus accept business considerations as appropriate criteria for economic and engineering decision-making. Although some of them voted for Socialist Party candidates in the 1970s, almost all of the *ingénieurs* (but not the *techniciens supérieurs*) opposed the nationalization of more private companies. Those few who expressed more radical views of capitalism held those views before starting full-time work. In this respect, French engineers resemble engineers in other advanced societies.

Similarly, despite resenting certain authoritarian managers, these engineers also value the authority of office for its positive uses in doing their jobs and advancing their careers. Although they complained about not being consulted more in company decision-making, these complaints were equally prevalent at both the traditional and high technology firms. Moreover, they are broadly French in character, reflecting the managerial role of *Monsieur l'ingénieur* in the past, the recent growth in the average size of French companies, the long-recognized tendency towards centralized and secretive management in French organizations, and the widespread demands for more "participation" that surfaced in France in the 1960s and 1970s.

Most importantly, all French engineers share with each other participation in a distinctively national system of technical employment and industrial politics. Thus, while their work and market situations may vary, the opportunities and constraints they face when thinking about changing category, function, or firm, or when strategizing to improve the rewards accruing to their current position, are strongly shaped by the role of the *grandes écoles*, the *polyvalence* of engineering education, the law protecting the title of *ingénieur diplomé*, the place of *autodidactes*, and *cadre* barriers and institutions. Chapter 9 argued that one of the effects of this system is an undermining of occupational solidarities.

To understand the distinctively national structure of technical careers in contemporary France, it is helpful to consider the needs of French industry in the nineteenth century, the human resources available for addressing those needs, and the institutional and ideological context that shaped the adaptation of needs and resources to one another.

Most firms were small, family-owned, and owner-managed. In small firms, perhaps especially in Catholic cultures, management tends to be paternalistic. It certainly was in France, partly for reasons of labor shortages in a country marked by widespread peasant land ownership and low population growth (Stearns, 1978; Reid, 1985). Small, paternalistic firms tend to strenuously resist collective bargaining by workers, partly because they cannot afford concessions to labor, partly because owner–managers tend to take any challenge to patrimonial authority personally. In France – a "low trust" culture in the widely respected view of Crozier (1964) – the memories of popular upheavals, from the Great Revolution of 1789 to the Paris Commune of 1871, made management especially suspicious of workers and concerned about controlling them. That the labor movement, organizationally weak and confronted by managerial intransigence and legal obstacles, advocated socialist or anarcho-syndicalist revolution further exacerbated class tensions in French industry. In such conditions, supervision is important and problematic, and owners seek as managers men who are both entirely trustworthy and also capable of legitimating their authority over workers by means of their own knowledge and hard work.

Small industrial firms also need technical assistance, but often can only support one engineer. That engineer must therefore be a generalist. More important for nineteenth-century France, however, is that there was a great shortage of highly skilled craftsmen from whom owners might have recruited artisan–engineers. This is not to adopt Dore's (1973) argument for Japan that late industrialization rendered the gap between artisanal and industrial skills so great that special schools were needed to provide technical training. Rather, as in Cole's (1979) persuasive critique of Dore's logic of late industrialization, it is to stress the pre-existing organization of labor and struggles over it specific to a single society. In short, French industrialists needed technical generalists and trustworthy, respectable managers, and had to look outside the firm to find them.

Outside the firm were occupations, classes, institutions, and ideas that shaped management's options and strategies. In the early nineteenth century there were the highly respected military and state engineers – technical officers who were graduates of elite state schools and who supervised as well as advised, but most of them disdained work in industry. There was also a general cultural tendency to rank the theoretical over the practical and the generalist over the specialist, and there were influential currents of Enlightenment ideals about reason and progress. Chapter 4 argued that in this context, an emerging technocratic ideology conceived of the engineer as the ideal manager of industrial progress.

If such traditional state engineering schools as the Polytechnique and the Ecole des Mines would not produce industrial engineers, a new school, the Ecole Centrale, would. Chapter 4 described the establishment of this school on the Polytechnique model, its effort to maintain respectability through *la science industrielle*, its legitimating functions for industrial managers, and its enormous influence on the subsequent defining and educating of all French engineers. A major reason for its success is that its rigorously selected, middle-class graduates came close to meeting the needs of French industrialists for both trustworthy managers and school-trained technical generalists. Thus developed the distinctive and enduring French practices of recruiting engineers from elite *grandes écoles*, defining them as prestigious technical officers, and favoring them for positions in top management. Once established, these practices induced ambitious young people to choose engineering as a route to higher management, and higher managers returned to this reliable source for the ambitious and trustworthy engineers and managers they needed. That is, there emerged processes of social reproduction linked to the different resources available to different social strata.

As firms grew larger in the twentieth century, other administrative workers emerged just below *Monsieur l'ingénieur*, while owners and top managers began to constitute an elite from which practicing engineers were increasingly excluded. Similar developments occurred throughout the industrial world, but only in France did these new middle-level occupations group together as *cadres*.

Chapter 7 discussed some of the distinctively French historical conditions that led to the emergence in France in the 1930s of a broad stratun of professional, administrative, and managerial employees: *cadres*. This loosely defined, multi-occupational stratum then absorbed the prestige, identities and organizations of engineers. *Cadres* also obtained some legal standing and national pension benefits, but not the right to control their own membership.

With the law of July 10, 1934, however, engineers did gain legal protection of the title, *ingénieur diplomé*, and engineering associations have had considerable influence on the state policies governing the subsequent expansions of engineering schools. Thus, there has emerged the crucial distinction between *ingénieurs diplomés* and *ingénieurs autodidactes* without the disappearance of either one. In view of the inability of the *ingénieurs diplomés* to obtain a legal monopoly over the practice of engineering, management has remained free to promote bright mechanics and technicians to the position of *Ingénieur* I, etc., an official position in the industry-wide collective bargaining agreements on job and salary hierarchies. Yet, the *ingénieurs diplomés* have been able to retain their relatively favorable position on the labor market. The result is France's unique structure of technical careers.

Post-industrial society: Convergence or divergence

The key recent developments have been those in the labor market. The rise of the high technology industries has been significant, but mainly in the demand it has created for technical workers, not the effects on the organization of technical work. More important has been the way this new demand has been met, a way that reflects France's distinctive institutions for producing and defining *techniciens* and *ingénieurs* but a way that is also leading to new tensions within French industry. Of particular significance has been the political compromise that resulted in the disproportionate expansion of *petite* engineering schools, by comparison to both the elite schools and the training programs for *techniciens supérieurs*. The elite schools and their politically powerful alumni associations were able to resist attempts at expanding the elite *grandes écoles*; the IUTs and *Lycées techniques* attempted to greatly expand their *technicien supérieur* programs, but fell far short of their goals, probably because many were seen as offering what Boudon (1977) aptly calls a "bad bargain." What was bad about the bargain for *techniciens supérieurs* was the probable career pay-off by comparison with that for graduates of the only slightly longer and more competitive *petite* engineering schools. The *technicien supérieur* starts his or her career near the top of the stratum between *cadres* and workers in France, but then faces the infamous *cadre* barrier and, if successful in surmounting that barrier, life as a relatively immobile *ingénieur autodidacte*. The graduate of a *petite* engineering school possesses a legally protected title, *ingénieur diplomé*, a title that suggests trustworthiness and assures initial hiring as a *cadre* in any large firm. The disappointments and resentments found among many French *ingénieurs diplomés* today are best understood by reference to the changes in their career and authority situations.

To the extent that science-based industries do require more experts who must be trusted, this problem of an expanding technical labor force is inherent in advanced industrial society. The consequences, however, depend considerably on the preexisting social definition and career structures for engineers, and on somewhat autonomous developments in the social production of engineers. Where engineers have constituted an occupation particularly favored for the recruitment of top managers, it is inevitable that a major expansion in their ranks will entail a diminution in the career possibilities of the average engineer.[1] In France, the problem has been compounded by industrial concentration and by a growing preference for top managers whose training and previous experience have been in finance or marketing rather than engineering.

Where engineers have traditionally served as line (rather than staff) officers in factories simmering with worker–management tensions, it is predictable that expansion outside the traditional blue-collar industries will affect relations with manual workers. Where young engineers are accustomed to collaborating closely with local management, a much faster increase in the number of young engineers than in the number of local managers is bound to diminish consultation by managers with engineers.

In emphasizing the distinctively French character of the career frustrations of French engineers, it is interesting to note also the distinctively French mechanisms for containing such frustrations. In France, engineering careers are more determined by educational credentials than they are in Britain or the United States, and educational credentials are allocated by means of a rather strict and age-conscious tracking system. The linking of engineering careers to the prestige of engineers' schools means that the disappointing realities of the careers of many are faced at an early age, not mid-career. With fairly successful claims of meritocratic justification, this system tends to "cool out" the disappointed and frustrated aspirants to traditional *grande école* careers.

To argue for the importance of such French institutions is to argue against theories of work organization and politics that emphasize the logic of either capitalism or industrialism. It is not, however, to advocate a theory of national culture, or even to contend that the effects of industrial developments on workers' ideologies are strongly conditioned by preexisting national values and national patterns of social organization. The French institutions that I have stressed are not global characteristics of French culture any more than they are characteristics of the work done by French engineers. Rather they express the social definition, production, and organization of technical workers in French society. If the argument presented here is persuasive, one implication is that future research should focus on this intermediate level of analysis, a level between that of the job and that of capitalism or industrialism.

Finally, it is important to indicate the limits of this approach. It is particularly applicable to middle-class, "knowledge" occupations. These are the occupations in which career structures are most significant and for which training outside the work organization is increasingly required. It is in and through educational structures and career patterns that national differences are manifest. If the advanced societies are indeed becoming knowledge-based, it may well be that modern industries are becoming increasingly dependent on institutions that are nationally distinctive. That is, contrary to the logic of industrialism and its implications for convergence among the advanced societies, the rise of knowledge-based occupations may be leading to a certain amount of divergence in institutional arrangements and political consequences among the nations which comprise the world of advanced industrial society.

Appendix
Interview schedule (English translation)

A Work section

1 To begin with can you briefly describe to me your current job? Typical day? Responsibilities?

2 In doing this work, do you most often use knowledge learned at school, or knowledge obtained in company training programs, or rather that acquired on the job? PROBE ON NATURE OF KNOWLEDGE NEEDED.

3 When a technical problem arises, how do you go about solving it?

4 Do you use your technical skills as much as you would like?

5 Do you remember how you happened to go into engineering in the first place?

6 Do you ever find yourself wishing you had chosen another occupation? IF YES: Which one? Why? IF NO: If you had to quit engineering all of a sudden and you could choose any other occupation, what would be your choice? (Why?)

7 LOOK AT WORK HISTORY ON WRITTEN FORM; IF APPROPRIATE: I see that you began your career at [another company]. Why did you leave _____ for _____?

8 How did you come to be working for [CURRENT COMPANY]?

9 IF ANY CHANGE IN ENGINEERING FUNCTIONS OR SPECIALTIES, ASK FOR EACH: I see you've changed from _____ to _____ since coming to [CURRENT COMPANY]. Why?

10 Is there anything about your previous jobs that you miss now?

11 Considering all the activities that your current job entails, which are those that give you the most satisfaction and which do you find most disagreeable?

12 Do you plan to remain at [CURRENT COMPANY]? If you were to look for another job, do you think it would be easy to find one?

13 In the past what sorts of things have been most valuable to you in getting the kind of job you wanted? IF NECESSARY: For example, your diploma, your past experience, a particular skill, personal contacts . . . ?

14 Have you ever been unemployed? IF YES: When was that and what were the circumstances?

15 How secure do you feel your present job is? Why? IF SEE JOB AS SECURE: Can you imagine any circumstances in which you would be in danger of losing your job?

16 When you think about the way your career has gone up to now, is there anything about which you are especially pleased or disappointed? [PROBE FOR EXPECTATIONS]

17 Looking to the future, in what sort of position would you like to be in five or ten years from now?

18 How do you view your chances of success? On what would you say they will depend?

19 In general, to have a good career, is it preferable to be a specialist or rather more of a generalist?

20 Do you think there are any jobs here that are so specialized that they are no longer interesting for an engineer?

21 And what about your job; is it too routine, pretty varied, or sometimes one, sometimes the other?

22 Does it ever happen that you can't do a job as well technically as you would like, for reasons of time, finances, commercial considerations, etc.?

23 Do you pretty much get to work on the kinds of projects and problems that interest you?

24 Do you think that, in order to do your current job better, it would be a good idea to extend your authority, your responsibilities or the range of your activities?

25 Are you ever obliged to do things you disagree with? PROBE ON REACTIONS, ATTITUDES TOWARDS AUTHORITY.

26 In your opinion, is it more difficult to exercise authority today than it was in the past?

27 In the type of work you do, do you think it's necessary to have a supervisor?

28 Do you know how personnel evaluations are done here? To what extent are you informed of your own evaluation?

29 If your supervisor isn't satisfied with your work or the work of someone like you, what is likely to happen?

30 In general, would you say the control of your work and your results is relatively strict, or rather relaxed, or what?

31 Are there some decisions on which you feel that you and the other *cadres* here ought to be more consulted or better informed? Which ones? Why? What about the workers or technicians?

32 Do you think top management here does all it can for the employees?

33 And yourself? Do you ever do things for the company which you are not obligated to do? What sorts of things? Do you feel the exchange with the company is a fair one in your case?

34 In your opinion, would it be better if the chief executive officer of [CURRENT COMPANY] were someone with training and experience in engineering, sales, finances, or administration?

35 What do you think of the salaries engineers earn at [CURRENT COMPANY]? What about the salary differentials? (among engineers, all employees) And how about your own salary, is it fair?

36 What's your opinion of average engineering salaries in France, compared to the average salaries of free professionals?

37 In your opinion, what sorts of things should be taken into consideration in setting the salary of a *cadre*?

38 In the military service, people often talk about the "morale" of the troops. Does it make sense to talk about the morale of *cadres* at [CURRENT COMPANY], and if so, how would you describe it at present?

39 What do you think of the relations between labor and management here? Do you personally feel closer to management, the technicians (IN THE CASE OF TECHNICIANS: workers), or neither one nor the other?

40 Returning to the *cadres* themselves, do they need to be organized? Why? IF YES: What ought the program and tactics of such an organization be? [PROBE FOR COLLECTIVE BARGAINING, STRIKES]

41 If it's not too indiscreet, what list do you normally vote for in company elections? Why?

42 Do you belong to a union? IF YES: Which one? Have you ever belonged? IF YES: Which? IF CHANGED: Why?

43 Do you belong to any professional associations, scientific societies, or alumni associations?

44 Have you ever written any scientific or technical articles?

B Family and politics section

LOOK AT PRELIMINARY QUESTIONNAIRE RESPONSES ON FAMILY.

1   I see you live at _____. Why did you choose to live there?

2   Are you satisfied with that neighborhood; this region?

3   IF WIFE IS NOT EMPLOYED: I see that your wife is not employed. Was she before your marriage (or the birth of your children?) IF YES: What did she do?

4   IF WIFE IS EMPLOYED: I see that your wife works as a _____. What do you think of her working? Does it cause any problems regarding the organization of your home life? How do you deal with these problems?

5   IF HAS CHILDREN: Do all your children live at home with you? What sort of school do they attend? IF BEYOND SCHOOL: What sort of schooling did they get? What are they doing now?

6   In the world today, what is the most important thing that parents should teach their children?

7   All parents seem to have certain hopes for their children. What about yours for your children?

8   In general, what are the consequences of your work life for your private and family life?

9   What do you do for leisure?

10 Can you briefly describe the two or three persons you consider your best friends? PROBE FOR WHERE MET, EDUCATION, JOB.

11 Among the people you work with, how many do you regard as friends?

12 Do you know your neighbors? IF YES: How well?

13 Do you belong to any clubs or associations, such as sports, or cultural, or civic, or religious, or charitable ones?

14 What, if any, is the daily newspaper you prefer to read? Do you subscribe? Read it regularly?

15 And what about weekly or monthly magazines?

16 Are you interested in politics?

17 Have you ever happened to do more than simply vote? For example: attend political meetings, participate in an electoral campaign, sign a petition, write a letter, or participate in a political demonstration?

18 What is or was the dominant political tradition in your family (parents)? For example: Gaullist, Socialist?

19 And you? Generally speaking, how would you characterize your own politics? To what, if any, party do you feel closest? Always that one? Are you active in or a member of any party or political movement?

20 Did you vote in the last election (1978 legislature)? IF YES: For what candidate did you vote, that is for the candidate of which party on the first ballot? And in the second round?

21 Did you vote in the Presidential elections of 1974? Do you recall for whom? *Municipales* (1977)?

22 In your opinion, are there any political parties, or groups, or people in France who have too much power or influence?

23 Do you think it would be better if in the major cities the mayors were not politicians? IF NECESSARY: better if they were managers?

24 Can you influence government policy in those areas of greatest concern to you? Would it be useful to see your representative in parliament?

25 What about taxes? Do you think the current tax system is fair?

26 And what about the manner in which the tax revenues get spent?

27 There's a lot of discussion these days about nationalizations. What's your opinion?

28 There have been a lot of debates on the subject of workers' control. Would you say that it's a matter of an insignificant myth, an interesting idea that's worth specifying, a form of organization that we should try to apply now, a dangerous idea, or what?

29 What do you think about the French nuclear energy program? PROBE ON FAITH IN TECHNICAL SOLUTIONS.

30 One can describe the differences that exist in a people in numerous ways. In your opinion, what is the most important criterion, among the following, for distinguishing people in France: age, region, occupation, religion, education, salary, life-style?

31 In your opinion, within French society, is a *cadre* (*technicien*) like yourself closer to manual workers, office workers, top managers, civil servants, or free professionals?

32 I'm particularly interested in social categories. According to you, is the word "class" useful, and if so, how many would you say there are in France, and how would you describe them?

IF NECESSARY:

33 In which class would you say you belong, you and your family?

34 In order to determine what class a person belongs to, what's most important to consider?
   IF NECESSARY: for example: education, income, life-style...

35 Here's a card which lists several categories of annual income. Could you tell me the letter which corresponds to your category? And for your entire household in 1976?

36 One talks a good deal in France about equality. Could you tell me what this term means to you?

37 One talks a lot about liberty, too. What are, for you, the most important liberties? Are they threatened in any way today?

38 I would like to ask you the nationality of your parents? And your grandparents?

39 And the religion of your parents? Were they practicing _____? Is it the same for you?

40 We've talked about your work life and your family life. When you think about the whole of your life, what are the things or types of activities in which you find the most satisfaction?

Notes

1 Technical workers in the advanced societies

1 See Grunberg and Mouriaux (1979: 44–45) for French figures.
2 For summary treatments of this literature, see the standard texts by Hall (1975), Montagna (1977), and Ritzer (1986). For dated but still valuable introductions to the field, see Caplow (1954) and Hughes (1958).
3 On the characteristics of professions, see Carr-Saunders and Wilson (1944), Parsons (1949), Greenwood (1957), Gross (1958), Goode (1957, 1960), Wilensky (1964), Vollmer and Mills (1966), and Freidson (1973b).
4 Some companies encourage and reward a cosmopolitan orientation among their engineers, thus reducing the potential for role strain; see Wilensky (1964), Glaser (1963), and Marcson (1969).
5 Meiksins (1982) has made a similar argument, but more recently (Meiksins and Watson, 1987) he has qualified it considerably, criticizing Derber's claims about ideological proletarianization in particular.
6 One reason the term *cadres* is so difficult to translate is that its meaning in French is imprecise and evolving. Traditionally, French government agencies and labor unions have used a broad definition that treated technicians, primary school teachers, social workers, and office supervisors as *cadres moyens*, reserving *cadres supérieurs* for *ingénieurs* and *cadres administratifs supérieurs* (accountants, sales executives, general managers). Industry on the other hand treats the *cadres moyens* as non-*cadres*. Since 1982, the French census bureau has moved towards the industrial definition, the definition I use in this study and that makes more sociological sense. For useful discussions see Groux (1983: 9–14), Grunberg and Mouriaux (1979: 10–14), and Boltanski (1983).
7 For valuable critiques of Braverman, see Burawoy (1978), Form (1980), Hill (1981), Larson (1980), Rubery (1987), Stark (1980), and the essays in Wood (1982).
8 For these "technocratic consciousness" as legitimation arguments see especially Marcuse (1964: ch. 6) and Habermas (1970: ch. 6). For an excellent critique of them, see Gouldner (1976).

9 Naville (1963) is sometimes regarded as a new working class theorist, but he focuses exclusively on automated plants so we ignore his analysis in the following discussion. For a summary of Naville's argument, see Gallie (1978: 21–25).

10 The "debate" began in *Arguments* in 1959; Gorz's major contribution was, in English, *Strategy for Labor* (1967). See also Belleville (1963).

11 See also his more elaborate critique, "The New Class: A Muddled Concept" in Bruce-Briggs (1981).

12 I have not discussed several good case studies, for much of the in-depth interviewing of French engineers and *cadres* has been done within the framework of organizational sociology, thus obscuring issues of class and politics. In the area of organizational sociology, see the series of studies by Maurice and his colleagues at the Laboratoire d'Economie et Sociologie du Travail (LEST), especially Maurice (1977); the studies of Crozier and his associates at the Centre de Sociologie des Organisations (CSO), especially Crozier (1964), Bachy et al. (1974), Sainsaulieu (1972), and Depuy and Martin (1977); and the study by Karpick (1978) at the Centre de Sociologie de l'Innovation at the Ecole des Mines.

13 For criticisms of Goldthorpe et al. (1969), see Mackenzie (1974), Blackburn and Mann (1979), Gruenberg (1980), and Hill (1981).

2 The companies: PAMPCO and TELECO

1 Freidson includes teachers and physicians, Galbraith sales executives, and Gorz students.

2 In the electrical construction industries in general, 30 percent of *cadres* are *cadres administratifs*, *commercial*, or *de la direction*, the rest being technical, but in professional electronics the technical component is certainly higher, just as the total percentage of *ingénieurs et cadres* at TELECO is 12.6 percent vs. 9 percent for the electrical construction industry in general (UIMM, 1977: 25–26).

3 The actual names of interviewees were chosen, using a random number table, from among all who remained eligible after application of the various sampling criteria. In some cases, it was necessary to include all eligible engineers. Fortunately, only one engineer at each company refused to be interviewed.

4 I conducted the majority of them, but my wife, Liliane Floge, conducted a good many, and two other assistants conducted several in the Paris region.

5 The interviews were tape-recorded in all but a few cases, but the tapes have never been transcribed and their use has been confined to confirming written notes and obtaining material for quotation. The coding scheme was devised in light of the types and ranges of answers given, after which the interview data were key-punched for computer analysis. The data analysis consisted of a combination of statistical analysis and the qualitative examination of the written or recorded responses to questions.

3 Pierre in his own words

1 This transcription and translation were done by Mrs. Nicole M. Polayes, a professional translator. The central criterion for the subsequent editing was the relevance of the interviewees' remarks to the topics addressed in the book.

4 The social meaning of technical knowledge

1 In a personal communication, André Grelon properly points out that the *poly-techniciens* have long followed their highly theoretical education at the Poly-technique by two years of much more concrete training in *écoles d'application*. I would simply add that much of this more practical training was more relevant to civil engineering projects for the army or other state agencies than to industrial engineering.

2 Grelon points out that Arts et Métiers was not the only model for subsequent engineering schools. The demand for specialized schools was greatest from 1900 to 1940, but after World War II there arose a fresh wave of interest in generalists. This gave rise to the Instituts Nationaux des Sciences Appliqués. For more on this subject, see Grelon (1983).

3 Boudon's figures are for registered students in the two year programs. I have divided them in half for 1961 but conservatively assumed a higher attrition rate in 1975, during which year there were 87,900 students in the IUTs and STS (*sections de techniciens supérieurs*).

4 To be sure, the distinction between research and design is questionable for elec-tronics. However, even if we combine these two functions, the resulting 42 per-cent is still below the 47 percent for all industries (including banking, but also government and academic research), although well above the 37 percent for the mechanical industries and 26 percent for mining and metallurgy.

5 Groux (1983: 24–27) argues that the proportion of all French *cadres* and *ingénieurs* constituted by *autodidactes* is declining with the growth of both the total numbers of *cadres* and *ingénieurs* and the continuation of credentials inflation. His evidence is that the proportion of *ingénieurs* having at least a *baccalauréat* increased from 67 percent in 1968 to 70 percent for men and 73 per-cent for women in 1981. But the issue is not the proportion having the *Bac* but rather the proportion having a *diplôme d'ingénieur*. No doubt the credentials inflation in more formal education makes today's *autodidactes* more highly schooled than their predecessors. But to the extent that Bell's point about theory implies a growing need for a kind of training that only professional schools can provide, the issue is one of diplomas from engineering schools. According to UIMM (1977: 19), "the pergentages of *cadres techniques diplomés* have varied relatively little from 1962 (51.8%), to 1970 (50.19%) to 1975 (49.8%)." UIMM gives the figures for all *ingénieurs et cadres diplomés* as a percentage of all *ingénieurs et cadres*, in which case the increase is from 36 to 40 percent. I have used the tables on distributions by function for both *nondiplomés* and all *ingénieurs et cadres* to separate out the figures for *ingénieurs* alone, assuming that all who work in technical functions (*technico-commerciaux*; *laboratoires*, *études et recherches*; *fabrication et entretien*; and *méthodes*, *contrôle et essais*; not *commerciaux*, *administration*, *direction* [higher management] or *divers*) are *ingénieurs*. See UIMM (1977: 12, 22, 59).

5 The organization and experience of technical work

1 Concerning manual workers, see Paul Adler (1986); David Halle (1984); Larry Hirschhorn (1984); Kenneth Spenner (1983); and Stephen Wood (1982). Concerning clerical workers, see especially K. Prandy, A. Stewart, and R. Blackburn (1982).

2 Although there is no definitive history of the engineer in France, there is a rapidly growing historical literature that offers many insights into the work and authority of nineteenth- and early twentieth-century engineers. See especially the recent volumes edited by Grelon (1984, 1986) and Thépot (1985), and, in English, the works of Fridenson (1985) on the car industry and Reid (1985) on mining and metal-working.

3 As Benguigui et al. (1978) note in their fine study of *cadres*, there are surprisingly few ethnographic accounts of such work in the sociological literature.

4 The categories in these time-budgets are a slight modification of those used by Benguigui et al. (1978). I am especially indebted to Dominique Monjardet for valuable assistance with this phase of the project.

5 This analysis owes much to a conversation with Robert Zussman.

6 The difference in the proportions expecting to leave one day reflected the superior external market situation of *ingénieurs* in the electronics industry. PAMPCO was losing money at the time of the interviews, and traditional iron and steel products throughout the advanced industrial societies were suffering from intense competition from imports.

7 Although focusing on "objective" rather than "subjective" data, Smith (1987: 156–157) reports a somewhat similar situation among the draftsmen, programmers, and other "established" technician occupations in the British aerospace plant he studied: "The skilled nature and 'legitimacy' of the established occupations were not under constant evaluation . . . The concern for quality not quantity, the sense of craft independence and the respect for engineering skill meant that these technical workers' daily routines were not bound by the fetters of quantifying and measuring the results of the labour of directly productive manual workers." Elsewhere (p. 78) he writes that "The main area of blocked mobility has been within the categories of draughtsmen and other technicians, as the rise of degree-holding (qualified) engineers has created a graduate ceiling that those without degrees cannot break through."

8 Smith (1987: 157) reports that in Britain, "the expansion in technical work in the 1970s was taking place at the higher end, amongst graduate engineers and technologists."

6 Autonomy and authority on the job

1 It is possible that the interviewees were cautious in expressing themselves on this subject, for the interviewing was done at work with the permission of the company. However, the unions also endorsed the research, and judging from the frankness with which many interviewees expressed themselves on other controversial matters – recall the interview with Pierre – significant misrepresentation of attitudes here seems unlikely.

2 See Zussman (1985) for a similar analysis of American engineers.

3 Littler (1982: 21–45) offers some useful introductory remarks on the development of Japanese and British industrial paternalism. With respect to Japan, he
notes (p. 38): "In general, many of the welfare practices of pre-war Japan have
become institutionalized as a matter of right. Paternalism has become impersonal
and contractual and we can see the crystallization of a new model of work organization, namely *bureaucratic paternalism*." Although more rooted in labor law
than in labor contracts, modern French industry also reveals a combination of
bureaucratic and paternalistic management. However, the French combination
leaves management with much greater discretion over such matters as pay
bonuses, while increasing the social distance between workers and managers
through such practices as recruiting *cadres* from among highly educated outsiders. The result is a much lower degree of social integration of workers in
French companies.

4 If the sales engineers are included, the figure at TELECO climbs to 42 percent,
for all seven of them complained, often bitterly, about having to clear everything
with their *chef*. It seems appropriate to leave them out, since there appears to be
no relationship between the technological level of an industry and the proportion
of its engineers in technical sales work (UIMM, 1977: 26). The other functional
differences were small; only maintenance stood out; 57 percent of maintenance
engineers complained, as opposed to 38 percent of design and production
engineers, 32 percent of research engineers, and 18 percent of methods
engineers. To the extent that research engineers are under-represented, the
difference between the over-supervision figures would be even larger than the
45–34 percent one presented above.

5 In France participation often connotes formal, institutionalized procedures for
participation by personnel in decision-making.

6 Hostility towards workers' unions is strongly associated with the number of subordinates an engineer has, and the latter varies widely by function. While
80 percent of production and 50 percent of maintenance engineers have 10 or
more subordinates, 17 percent or fewer of those in all other functions do. In that
the company samples over-represent production and maintenance engineers at
TELECO, and research and design engineers at PAMPCO, the actual inter-firm
(and inter-industry) differences in feelings about unions and the difficulties of
supervising are even more pronounced than indicated here.

7 Studies of American (Zussman, 1985) and British (Whalley, 1986; Smith, 1987)
technical workers report similar findings. Thus, Smith (1987: 244) writes:
"Whether there existed an 'us' and 'them' situation in the office, or a situation of
mutual co-operation, was down to how the manager *acted*, whether he took an
interest in the job, was a disciplinarian or indifferent and a recluse. Their assess-
ment of management in general, was considerably influenced by the character of
the particular office manager."

8 Zussman (1985), Whalley (1986), and Smith (1987) report similar findings. For
example, Smith (1987: 162) writes: "The designers saw technical change largely
in straight capitalist terms of staying in the field, by keeping up with or ahead of
the firm's competitors. 'Competition', 'profit', 'efficiency' were part of the

language they used to explain the necessity of technical change. There was no sense of engineering and capitalist logics being oppositional alternatives."

9 While the tradition of *Monsieur l'ingénieur* makes French engineers especially vulnerable to such feelings of lost authority, they are found even among some British draftsmen. Thus, Smith (1987: 94) reports that several draftsmen in their fifties bemoaned this loss of authority, Eric Ham for example: "I can remember when the draughtsman went into the shops. I mean his word was absolute law and the men of the shop floor used to say 'Oh, there's a draughtsman coming down.'"

7 Labour markets and career experiences

1 The FASFID surveys give age distributions, but the low response rates – 21.8 percent in 1980 – make these FASFID figures particularly unreliable.

2 According to Levy-Leboyer's historical analysis (1979: 152) of company directors, the percentage having graduated from engineering schools other than the Polytechnique has declined from 50 percent in 1919 and 55 percent in 1929 to 31 percent in 1973. (Graduates of the Polytechnique accounted for another 20 percent in both 1929 and 1973.) Yet, while these figures are revealing about the educational background of top management, they prove nothing about the chances of engineering graduates to reach top management.

3 See Boltanski (1987: ch. 2) for a richly detailed and sophisticated discussion.

4 The Parodi Index is being phased out in favor of a simplified scale, but was still in use at the time of this research.

5 It is interesting to compare a more Marxist analysis that is sensitive to historically developed national differences in technical education and careers, but emphasizes "capital" at the national level rather than the autonomy of individual firms. Thus Smith (1987: 163) writes: "While American corporations structured the education system to produce specialists and corporate career structures to turn them into coordinators, British capital left the engineer a specialist and drew off elite educational institutions for global functionaries."

6 See the 1971 survey by L'Idres-Marketing, reported in *Enterprises* (December, 1971) and referred to by Boltanski (1982: 427).

7 See also Haller et al. (1985).

8 The June issue of *L'Expansion*, 1978, commenting on its 10th edition of its annual special, "Le Prix des cadres," noted that ten years ago no one knew about Apec because "unemployment among cadres was unknown and unimaginable" (p. 157).

9 On the rise of the *petites écoles*, see also Gilpin (1968: 352).

10 French *cadres* often argue that their salaries are much less impressive after taxes and family allowances, etc., but Groux (1983: 50) presents figures for seven advanced countries that show high French salaries suffering the lowest reductions from taxes and other deductions.

11 Comparisons between companies are complicated by several factors, including differences in the ways each company calculates annual salary (including bonuses), classifies individuals, and summarizes data. Additional complexities arise from the different distributions of ages in each firm, and from regional wage

differentials. Fortunately, the engineers at both firms conduct very detailed salary surveys among themselves periodically, and they and the personnel offices of PAMPCO cooperated fully with this research.

12 · In a recent issue of *La Monde de l'Education* (May 1984) devoted to questions about engineering as a career, we find the following. "What defines an *ingénieur*, finally, is his position – infinitely variable – in a hierarchical complex, much more than his 'métier' or the nature of his function" (p. 32).

13 Thirty-one percent were neither disappointed nor pleased; 39 percent expressed disappointment.

14 Seventeen percent emphasized autonomy, 20 percent the content of the work, and the rest "other."

15 One personnel administrator claimed that the market was better for people like him than for engineers. He criticized the delays in the national education system's adjustments to market conditions, saying the result had been an oversupply of young *ingénieurs diplomés* and *techniciens*, especially in electronics, where demand was now leveling out.

16 Rank was scaled as follows: a *cadre I* (*débutant*) got a 3, a *cadre II* (*confirmé*) a 4, a *cadre IIIA* a 5, etc. Both companies followed France's fairly standardized system for grading positions.

17 There is some misplaced precision in these averages, for the salary question simply asked interviewees to indicate in which of several salary categories their own fell.

8 Trade unions and professional associations

1 On the history of French unionism in general, see the three volumes by Georges LeFranc (1950; 1967; 1969). On white-collar unionism, see Crozier (1966). On the history of collective organization among *cadres*, see Maurice (1971), Grunberg and Mouriaux (1979), Boltanski (1982; 1983), Thépot (1985: 217–228), and especially Descostes et al. (1984).

2 SPID: Syndicat professionnel des ingénieurs diplomés; USIF: Union des syndicats d'ingénieurs français; SIS: Syndicat des ingénieurs salariés.

3 The two other federations were the Groupement syndical des collaborateurs diplomés (GSCD), founded in 1937, and the Fédération des syndicats de chefs de service, agents de maîtrise et techniciens des industries métallurgiques et connexes en France, the result of a merger in 1937.

4 At the end of 1977, the FNIC changed its name to the Union des cadres et ingénieurs (UCI).

5 The multi-occupational, stratum-conscious organization of French workers goes back to the *ancien régime*. William Sewell points out that even the illegal *Campagnonnages* " incorporated journeymen of different trades in their itinerant search for work." That is, despite the power of the legal craft guilds and the moral commitment of journeymen to the craft community, the illegal brotherhoods of journeymen not only split crafts by strata – journeymen and masters – but also grouped together journeymen from several occupations (1980: 42).

6 Grunberg and Mouriaux conclude that there has been, contrary to Galbraith's predictions, a *growth* in the rate of *cadre* unionization. However, in view of

Boltanski's point about the strange rise in the use of the term *cadres* and the disproportionate growth of the numbers of salaried workers officially classified as *cadres*, one must question the validity of figures comparing *cadres* in one time period with *cadres* in another. In any case, the enormous differences in the types of *cadre* unions in France make any summary figures about unionization rates relatively meaningless.

7 It is important to bear in mind that like many French scholars, Grunberg and Mouriaux have defined *cadres* according to official government classifications, while I have adopted the much narrower definition observed within French industry. It is for this reason that *techniciens* are included in Grunberg and Mouriaux's sample.

8 There are two other representative bodies to which employees in French firms elect delegates, the delegates of the personnel (*délégués du personnel*) and the Union Section (*Section syndicale*), but the *comité d'entreprise* is more important than either of these. See Gallie (1978; 151–157) for an unusually good discussion of all three institutions.

9 On occasions the workers' unions cannot find a *cadre* to run as a candidate within Collège III. This was often the case at paternalistic PAMPCO; few *cadres* would want to and none would dare. Lest this be taken as a sufficient explanation for the observed differences between *ingénieurs* and *techniciens*, it should be added that the normal reaction to being denied preferred candidates is abstention. Yet, the abstention rate is far higher among the *techniciens* (Collège II) than among the *ingénieurs* (Collège III). It seems that either PAMPCO is exceptional in the lack of choice offered Collège III electors, or that *cadres* simply do not care, supporting the CGC in either case. Noteworthy in this regard is that the abstention rates in *métallurgie* and *électronique* are virtually the same.

10 The figure for PAMPCO does not include the few additional engineers who, when asked if they belonged to a professional or scientific society, said something like: "No, except for my nearly obligatory participation in L'Association technique de fonderie."

11 The Société des ingénieurs civils was founded by several *centraliens* as a professional society open to all engineers, even *autodidactes*, for the purpose of advancing the position of non-state engineers in France, as well as exchanging knowledge. Over time, however, it developed into a learned society, as its other functions, especially that of mutual aid, withered. The latter gap was filled by the progressive growth of friendly societies of the alumni of specific *grandes écoles*: Centrale in 1862, Mines in 1864, Polytechnique in 1865. "They all have as their main purpose mutual aid among engineers graduated from the same school, which quickly generated an esprit de corps which went against the federal discourse of its [SIC's] founders" (Jacomy, 1984: 213).

9 Social participation, politics, and class

1 Unfortunately, the question was worded to include independent professionals as well as *cadres*. It is clear from the answers, however, that most of the friends who were assigned this code were *cadres*.

2 In Ehrmann's words (1983: 105), "Neither education (except for the buildings

and for janitorial help) nor the police forces (except for the *garde champêtre*) are financed out of the local budget. This means that local government authorities have little control over these services."

3 The three elections were the parliamentary election of 1978, the municipal elections of 1977, and the presidential election of 1974. In the case of the parliamentary election, the engineers were asked who they voted for in each of the two *tours* or rounds of balloting. The first *tour* in French elections usually offers a choice among several candidates who represent France's wide variety of political parties and ideologies. Since it is rare for any one of those candidates to win a majority, there is usually a second, run-off *tour* that forces a choice between the two top vote gainers in the first *tour*. Although the municipal elections of 1977 were unusually politicized and the outcomes in the larger cities regarded as indicative of the left's gains in France, voting in the small towns where many of the interviewees lived was often as apolitical as it has been traditionally. (See the discussion on pp. 211–12.) Therefore, I have largely ignored the responses concerning *les municipales*, and not used this item in the construction of the voting scales.

4 Votes for candidates of the right in either *tour* of the 1978 *législatives* and for Giscard in the presidential election of 1974 counted one point; votes for candidates of the left and for Mitterrand counted two. Thus, those who consistently voted for the right achieved the lowest possible score on the scale, a 3, while those voting consistently for the left scored a 6. To simplify the presentation in the text, this scale has been reduced to a dichotomy in the tables, with those scoring 4 being merged with those scoring 3, and those scoring 5 with those scoring 6. Only 16 of the 102 respondents voted inconsistently, so this dichotomization has little effect on the picture presented.

5 In three different multiple classification analyses, each employing a slightly different set of independent variables that were associated with voting (age, rank, function, firm, etc.), the beta for parents' politics was the highest in its set, statistically significant at a level of .01.

6 The seven categories and the percentages at each firm (PAMPCO/TELECO) favoring each of these categories were: (1) Center-right *majorité* (no party named) (12/3); (2) RPR or Gaullist (15/7); (3) PR (Giscard), UDF, or Centrist (254/30); (4) Ambivalent, Center-left, or "Social-democrat of the German variety" (27/26); (5) Socialist, PS, or left but anti-Communist (12/21); (6) Communist (0/1); (7) Other, left, including PSU, ecologists, and extreme left (4/12); (8) None (2/0); (9) Other and no answer (4/0).

7 In many ways, the situation of *techniciens* is similar to that of the *ingénieurs diplomés* at TELECO, while the situation of *ingénieurs autodidactes* is closer to that of the *ingénieurs diplomés* at PAMPCO. With the *techniciens*, it is not their pay that upsets them so much as their position on the internal labor market as a result of the *cadre* barrier. In view of their formal credentials and the shortage of them, however, they are in a strong position in the external labor market. This combination seems calculated to generate both anger at the system and the sense of independence needed to express such feelings in union and individual politics. *Ingénieurs autodidactes*, by contrast, lack the formal credentials and youth

needed to find work as *cadres* elsewhere, so are dependent on their current employers. Yet, their careers have turned out very successfully for Frenchmen starting where they did, and they are both happy about that, impressed by what individual initiative and hard work can accomplish, and grateful to management for having recognized their contribution. The fact that many of them work in production and struggle to supervise workers may increase their support for right-wing parties, but the peculiarities of their market situation have been too long ignored.

8   This question was asked in a written questionnaire sent several months after the interviews. In that the response rate was only 50 percent at PAMPCO and 47 percent at TELECO, the date leave much to be desired. However, since the distributions of respondents are fairly representative of the larger sample, and since the pattern of responses to this question is quite pronounced, I have decided to include these findings.

9   To obtain the data on class images, I followed the open-ended interviewing techniques developed and used by Goldthorpe et al. (1969), Mackenzie (1973), and Low-Beer (1978). For a review of these procedures, see Bulmer (1975).

10   Gordon Craig (1955: 27) provides an interesting military example. In explaining Napoleon's battlefield success, he writes: "The destruction of the old regime and the granting of fundamental rights to all citizens had an immediate effect upon the constitution of the French Army . . . It was no longer necessary for the French to concentrate their forces in close array upon the battlefield, forbidding independent manoeuvre lest it lead to mass desertion. The French tirailleurs advanced in extended order, fighting, firing, and taking cover as individuals, and the Army gained immeasurably in tactical elasticity in consequence. Troops could, moreover, be trusted to forage for themselves, and it was now possible to divorce French units from the cumbersome supply trains and the dependence on magazines which restricted the mobility of the old model armies."

11   For whatever merits there are in these criticisms, I owe much to stimulating conversations with Marc Maurice and Dominique Monjardet.

12   The only *cadres* for whom Boltanski shows much sympathy are the *cadres autodidactes*, whom he views as neglected and exploited foreigners within the *cadre* class.

13   On this, see especially Holmwood and Stewart (1983), and their forthcoming book.

14   Comparable analyses of American class and politics are found in Janowitz (1978) and Halle (1984). Halle shows how affluent blue-collar workers blur the collar line away from work, drawing the class line harder and higher below them (than they do at work) but higher and softer above them. Similarly, *cadres* don't stress the *cadre* barrier as much as *techniciens* do at work, but call more attention to *la direction*, a distinction little noted by the *techniciens*, and one that leads *cadres* to seek some common ground with *techniciens* in defense of their common interests as employees. In the more fluid status game away from work, these same *cadres* seem to blur the distinction between *cadres* and top managers, but emphasize education and style of life in ways that distinguish them from all below them at work while assimilating to those above. There is a general logic here that operates

similarly all up and down the class system. However, there is also some convergence between the worlds of work and community at the top of the hierarchy, and there are important variations among nations in the degree and character of disjuncture.

10 Conclusions: technology, nations, and career structures

1 This yields the more general hypothesis that the effects of an occupation's expansion (in response to market demands) upon its power and status depend upon the occupation's previous position. To use Parkin's terminology (1979: chs. 4–6), expansion of an occupation that depends on "usurpationary closure" and tactics will increase its power, but expansion of a privileged occupation that relies on "exclusionary closure" will reduce each member's access to scarce values.

References

Abercrombie, N. and J. Urry. 1983. *Capital, Labour and the Middle Classes*. London: George Allen and Unwin

Adler, Paul. 1986. "Technology or Us?" *Socialist Review*, 85

Ahlstrom, Goran. 1982. *Engineers and Industrial Growth: Higher Technical Education and the Engineering Profession during the Nineteenth and Early Twentieth Centuries: France, Germany, Sweden and England*. London: Croom Helm

Almond, Gabriel and S. Verba. 1963. *The Civic Culture*. Princeton University Press

Alt, John. 1976. "Beyond Class: The Decline of Industrial Labor and Leisure." *Telos*, 28: 55–80

Althauser, Robert P. 1987. "Internal Labor Markets, U.S.A.: A Thematic Review." Paper presented at the American Sociological Association Annual Meeting, Chicago, IL, August 20

Ardagh, John. 1982. *France in the 1980s*. New York: Penguin Books

Aron, Raymond. 1968. *The Illusive Revolution*. New York: Praeger

Aronowitz, Stanley. 1979. "The Professional-Managerial Class or Middle Strata" in Pat Walker (ed.), *Between Labor and Capital*. Boston: South End Press

Bachy, J. P., F. Dupuy, D. Martin, and J. D. Reynaud. 1974. *Etude sur les cadres supérieurs du groupe SAG–PAMP* (pseudonym). Paris: A.D.S.S.A.

Bailyn, Lotte. 1980. *Living with Technology: Issues at Mid-Career*. Cambridge, MA: MIT Press

Balassa, Bela. 1981. "The French Economy under the Fifth Republic, 1958–1978." In W. Andrews and S. Hoffman (eds.), *The Impact of the Fifth Republic on France*, pp. 117–138. Albany: State University of New York Press

Baron, James N. 1984. "Organizational Perspectives on Stratification." *Annual Review of Sociology*, 10: 37–69

Barrier, Christiane. 1968. "Techniciens et grèves à l'Electricité de France." *Sociologie du Travail*, 1: 50–71

Baudelot, Christian, R. Establet and J. Malemort. 1974. *La Petite bourgeoisie en France*. Paris: François Maspéro

Bauer, M. and E. Cohen. 1979. *La Production sociale de la Grande Entreprise Multi-branche*. Paris: C.S.I./D.R.G.S.T.

1980. "Les Limites du pouvoir des cadres: l'organisation de la negotiation comme moyen d'exercice de la domination." *Sociologie du Travail*, no. 3

1982. "Les Limites du savoir des cadres: l'organisation savante comme moyen de déqualification." *Sociologie du Travail*, no. 4

Bell, Daniel. 1972. "Labor in the Post-Industrial Society." *Dissent* (Winter): 163–189

1973. *The Coming of Post-Industrial Society*. New York: Basic Books

1979. "The New Class: A Muddled Concept." In B. Bruge-Briggs (ed.), *The New Class?*, pp. 169–190. New York: McGraw Hill

Belleville, Pierre. 1963. *Une Nouvelle classe ouvrière*. Paris: Julliard

Benguigui, Georges. 1981. "La Selection des cadres." *Sociologie du Travail*, 3: 294–307

Benguigui, G., A. Griset, A. Jacob and D. Monjardet. 1978. *La Fonction d'encadrement*. Paris: La Documentation française

Benguigui, G. and D. Monjardet. 1968. "Profession ou corporation? Le cas d'une organisation d'ingénieurs." *Sociologie du Travail*, 3: 275–289

1984. "The CGC and the Ambiguous Position of the Middle Strata." In M. Kesselman (ed.), *The French Workers Movement*. New York: Columbia

Bennis, Warren G. and P. Slater. 1968. *The Temporary Society*. New York: Harper and Row

Blackburn, Robert and Michael Mann. 1979. *The Working Class in the Labour Market*. London: Macmillan

Blauner, Robert. 1964. *Freedom and Alienation*. University of Chicago Press

Boltanski, Luc. 1978. "Les Cadres autodidactes." *Actes de la Recherche en sciences sociales*, 22: 3–24

1982. *Les Cadres: la formation d'un groupe sociale*. Paris: Les Editions de Minuit

1983. "How a social group objectified itself: *Cadres* in France, 1936–1945." Paper presented at the Fourth International Conference of Europeanists, Washington, DC, October 13–15

1987. *The Making of a Class: Cadres in French Society*. Cambridge University Press

Boudon, Raymond. 1977. *Effets pervers et ordre social*. Paris: Presses Universitaires de France

Braverman, Harry. 1974. *Labor and Monopoly Capital*. New York: Monthly Review Press

Bruce-Briggs, B. (ed.). 1981. *The New Class?* New York: McGraw-Hill

Bulmer, Martin (ed.). 1975. *Working Class Images of Society*. London: Routledge and Kegan Paul

Burawoy, Michael. 1978. "Toward a Marxist Theory of the Labor Process: Braverman and Beyond." *Politics and Society*, 8 (3–4): 247–312

Burchell, Brendan, and Jill Rubery. 1987. "The experiences of individuals in the labour market: Determinants of their employment expectations and job satisfaction." Paper for the Ninth Conference of the International Working Party on Labour Market Segmentation (ILO), Turin, Italy, July, 1987

Burns, Tom and G. M. Stalker. 1961. *The Management of Innovation*. London: Tavistock

Calhoun, Daniel. 1960. *The American Civil Engineer*. Cambridge, MA: Harvard University Press

Calvert, Monte. 1967. *The Mechanical Engineer in America*. Baltimore: Johns Hopkins University Press

Capdevielle, Jacques et al. 1981. *France de gauche, vote à droite*. Paris: Presses de la Fondation Nationale des Sciences Politiques

Caplow, Theodore. 1954. *The Sociology of Work*. Minneapolis: University of Minnesota Press

Carchedi, Guglielmo. 1977. *On the Economic Identification of Social Classes*. London: Routledge and Kegan Paul

Carr-Saunders, A. M. and P. A. Wilson. 1944. "Professions." *Encyclopedia of the Social Sciences*. New York: Macmillan Company

Carter, Reginald. 1977. "Are the Work Values of Scientists and Engineers Different than Managers?" In M. R. Haug and J. Dofney (eds.), *Work and Technology*, pp. 125–140. Beverly Hills, CA: Sage Publications

Cheverny, Julien. 1967. *Les Cadres: essai sur les nouveaux proletaires*. Paris: Julliard

Cole, Robert. 1979. *Work, Mobility and Participation*. Berkeley: University of California Press

Collins, Randall. 1979. *The Credential Society: A Historical Sociology of Education and Stratification*. New York: Academic Press

Craig, Gordon A. 1955. *The Politics of the Prussian Army, 1640–1945*. New York: Oxford University Press

Crompton, Rosemary and John Gubbay. 1977. *Economy and Class Structure*. New York: St. Martin's Press

Crozier, Michel. 1964. *The Bureaucratic Phenomenon*. University of Chicago Press

   1966. "White-Collar Unionism – The Case of France." In A. Sturmthal (ed.), *White Collar Trade Unions*, pp. 91–126. Urbana: University of Illinois Press

Crozier, M., F. Dupuy and D. Martin. 1975. *Les Cadres et l'organisation*. Paris: Association pour le développement des sciences sociales appliquées. Cultural Sevices of the French Embassy

Day, Charles. 1978. "The Making of Mechanical Engineers in France: The Ecoles d'Arts et Métiers, 1803–1914." *French Historical Studies* (Spring)

Dejonghe, Etienne. 1985. "Ingénieurs et société dans les houillères du Nord-Pas-de-Calais, de la 'Belle Epoque' à nos jours." In Andre Thépot (ed.), *L'Ingénieur dans la société française*, pp. 173–189. Paris: Les Editions ouvrières

Derber, Charles. 1982. *Professionals as Workers: Mental Labor in Advanced Capitalism*. Boston: G. K. Hall and Co.

   1983. "Managing Professionals: Ideological Proletarianization and Post-Industrial Labor." *Theory and Society*, 12 (May): 309–341

Descostes, Marc et al. 1984. *Clefs pour une histoire du syndicalisme cadre*. Paris: Les Editions ouvrières

Doeringer, Peter B. and Michael Piore. 1971. *Internal Labor Markets and Manpower Analysis*. Lexington, MA: D. C. Heath and Co.

Dore, Ronald P. 1973. *British Factory–Japanese Factory: The Origins of National Diversity in Industrial Relations*. Berkeley: University of California Press

Dubin, Robert. 1956. "Industrial Workers' Worlds: A Study of the Central Life Interests of Industrial Workers." *Social Problems*, 3: 131–142

Dubois, Jean. 1978. "Des 'Héritiers', toujours." *L'Expansion*, no. 121: 122–123

Dubois, Pierre et al. 1971. *Grèves revendicatives ou grèves politiques?* Paris: Editions Anthropos

Dulong, Renaud. 1971. "Les Cadres et le mouvement ouvrier." In Pierre Dubois et al., *Grèves revendicatives ou grèves politiques?*, pp. 161–245. Paris: Editions Anthropos

Dupuy, François and Dominique Martin. 1977. *Jeux et enjeu de la participation*. Paris: CRESST

Durand, Claude. 1968. "Introduction", *Sociologie du Travail*, 10 (2).

   1971. "Ouvriers et techniciens en mai 1968." In Pierre Dubois et al., *Grèves revendicatives ou grèves politiques?*, pp. 7–159. Paris: Editions Anthropos

Durand, Claude and Michelle Durand. 1971. *De l'O.S. a l'ingénieur: carrière ou classe sociale*. Paris: Les Editions ouvrières

Durand, Michelle. 1972. "Professionalisation et allégeance chez les cadres et les techniciens." *Sociologie du Travail*, 14: 185–212

Durkheim, Emile. 1947. *The Division of Labor in Society*. Glencoe, IL: The Free Press

Ehrenreich, Barbara and John Ehrenreich. 1979. "The Professional-Managerial Class." In Pat Walker (ed.), *Between Labor and Capital*, pp. 5–45. Boston: South End Press

Ehrmann, Henry W. 1983. *Politics in France* (4th edn). Boston: Little Brown

Etzioni, Amitai. 1968. *The Active Society*. New York: The Free Press

FASFID (La Fédération des associations et sociétés françaises d'ingénieurs diplomés). 1968–81. *3ème–7ème enquête socio-économique sur la situation des ingénieurs diplomés*. Numéro spécial, *ID* 71

Fores, Michael. n.d. "On Engineers in Western Europe." Unpublished manuscript, Department of Industry, London

Form, William. 1980. "Resolving Ideological Issues on the Division of Labor." In H. Blalock (ed.), *Sociological Theory and Research: A Critical Appraisal*, pp. 140–155. New York: The Free Press

Fossati, Hélène and Gérard Said. 1983. *Evolution technologique et restructuration industrielle; l'enjeu d'une formation: les détenteurs de BTS-DUT A Solmer Fos/Mer*. Document LEST 83/14

Fox, Alan. 1974. *Beyond Contrast: Work, Power and Trust Relations*. London: Faber and Faber

Freidson, Eliot. 1973a. "Professionalization and the Organization of Middle-Class Labour in Postindustrial Society." In P. Halmos (ed.), *Professionalization and Social Change*, pp. 47–59. University of Keele, England: The Sociological Review Monograph, 20

   1973b. "Professions and the Occupational Principle." In E. Freidson (ed.), *The Professions and their Prospects*, pp. 19–38. Beverly Hills, CA: Sage Publications

1986. *Professional Powers: A Study of the Institutionalization of Formal Knowledge*. University of Chicago Press

Fridenson, Patrick. 1985. "Les Ingénieurs et cadres de l'automobile en France au XXème siècle." In P. Joutard and J. Lecuir (eds.), *Histoire sociale, sensibilités collectives et mentalités*, pp. 431–448. Paris: Presses Universitaires de France

Galbraith, John K. 1968. *The New Industrial State*. New York: Signet Books

Gallie, Duncan. 1978. *In Search of the New Working Class*. Cambridge University Press

Garaudy, Roger. 1970. *The Crisis in Communism: The Turning Point of Socialism*. New York: Grove Press

Geison, Gerald L. (ed.). 1984. *Professions and the French State, 1700–1914*. University of Pennsylvania Press

Giddens, Anthony. 1973. *The Class Structure of Advanced Societies*. London: Hutchinson University Library

Gilpin, Robert. 1968. *France in the Age of the Scientific State*. Princeton University Press

Gispin, C. W. R. 1985. "German Engineers and American Social Theory: Historical Perspectives on Professionalization." Paper presented at Fifth International Conference of the Council for European Studies, Washington, DC

Glaser, Barney. 1963. "The Local-Cosmopolitan Scientist." *American Journal of Sociology*, 69: 249–259

Glover, Ian A. 1983. "How the West was Lost? The Decline of Engineering and Manufacturing in Britain and the United States." Paper presented to the American Sociological Association Annual Meeting, Detroit, Michigan

Goldner, Fred and Richard Ritti. 1967. "Professionalization as Career Immobility." *American Journal of Sociology*, 72 (5): 489–502

Goldthorpe, John. 1982. "On the Service Class, its Formation and Future." In A. Giddens and G. Mackenzie (eds.), *Social Class and the Division of Labor*, pp. 162–185. Cambridge University Press

Goldthorpe, J., D. Lockwood, F. Bechhofer, and J. Platt. 1969. *The Affluent Worker in the Class Structure*. Cambridge University Press

Goode, William. 1957. "Community within a Community: The Professions." *American Sociological Review*, 22 (2): 194–200

1960. "Encroachment, Charlatanism, and the Emerging Profession: Psychology, Sociology, and Medicine." *American Sociological Review*, 25 (6): 902–926.

Gordon, Allan. 1981. *The Development of the Engineering Profession in Canada, 1918–1971*. M.A. Thesis, Carleton University, Ottawa, Ontario

Gorz, André. 1967. *Strategy for Labor*. Boston: Beacon Press

1976. "Technology, Technicians and Class Struggle." In A. Gorz (ed.), *The Division of Labour*, pp. 159–189. Atlantic Highlands, NJ: Humanities Press

Gouldner, Alvin. 1954. *Patterns of Industrial Bureaucracy*. New York: The Free Press

1976. *The Dialectic of Ideology and Technology*. London: Macmillan Press

1979. *The Future of Intellectuals and the Rise of the New Class*. New York: Seabury

Granick, David. 1972. *Managerial Comparisons in Four Developed Countries: France, Britain, United States and Russia*. Cambridge, MA: MIT Press

Greenwood, Ernest. 1957. "Attributes of a Profession." *Social Work*, 2 (3): 45–55

Grelon, André. 1982. "La Modele de l'école d'ingénieurs comme formation à la technologie et comme insertion dans la société." *Informations sur les Sciences Sociales*, 415: 719–773

  1983. *L'Education des cadres*. Paris: E.H.E.S.S. – Université Paris VII: thèse de 3ème cycle sous la direction de M. Paul-Henri Chombart de Lauwe

  1984. "Les Ingénieurs, encore." In *Culture Technique*, 12 (March): 11–17.

  1986. *Les Ingénieurs de la crise*. Paris: Editions de l'école des hautes études en sciences sociales

Gross, Edward. 1958. *Work and Society*. New York: Thomas Y. Crowell Co.

Groux, Guy. 1982. *Les Ingénieurs et cadres dans le mouvement ouvrier*. Paris: Conservatoire National des Arts et Métiers

  1983. *Les Cadres*. Paris: François Maspero

  1985. "Le Syndicalisme–cadres: ordre, régulation, rupture." *Société Française*, 14: 30–36

Gruenberg, Barry. 1980. "The Happy Worker: Determinants of Job Satisfaction." *American Journal of Sociology*, 86 (2): 247–271

Grunberg, Gérard and R. Mouriaux. 1979. *L'Univers politique et syndicale des cadres*. Paris: Presses de la Fondation Nationale des Sciences Politiques

Habermas, Jurgen. 1970. "Technology and Science as 'Ideology'." In J. Habermas, *Toward a Rational Society*, pp. 81–122. Boston: Beacon Press

Hall, Oswald. 1948. "The Stages of a Medical Career." *American Journal of Sociology*, 54: 243–253

Hall, Peter. 1986. *Governing the Economy: The Politics of State Intervention in Britain and France*. New York: Oxford University Press

Hall, Richard. 1975. *Occupations and Social Structure*. New York: McGraw-Hill

Halle, David. 1984. *America's Working Man*. University of Chicago Press

Haller, Max et al. 1985. "Patterns of Career Mobility and Structural Positions in Advanced Societies: A Comparison of Men in Austria, France, and the United States." *American Sociological Review*, 50 (5): 579–603

Halls, W. D. 1976. *Education, Culture and Politics in Modern France*. Oxford: Pergamon Press

  1979. "France: The Central Light of Reason." In Edmund King (ed.), *Other Schools and Ours*. New York: Holt, Rinehart and Winston

Hill, Stephen. 1981. *Competition and Control at Work: The New Industrial Sociology*. Cambridge, MA: MIT Press

Hirschhorn, Larry. 1984. *Beyond Mechanization*. Cambridge, MA: MIT Press

Hoffmann, Stanley et al. 1963. *In Search of France*. New York: Harper and Row

Hofstede, Geert H. 1980. *Culture's Consequences: International Differences in Work-Related Values*. Beverly Hills, CA: Sage Publications

Holmwood, J. M. and A. Stewart. 1983. "The Role of Contradictions in Modern Theories of Social Stratification." *Sociology*, 12 (2): 234–254

Hughes, Everett C. 1958. *Men and Their Work*. New York: The Free Press

Hutton, S. P. and P. A. Lawrence. 1981. *German Engineers: The Anatomy of a Profession*. London: Oxford University Press

Jacomy, Bruno. 1984. "A la recherche de sa mission: la société des ingénieurs civils."
    In A. Grélon (ed.), *Culture Technique*, no. 12 (March): 209–219
Janowitz, Morris. 1978. *The Last Half Century: Societal Change and Politics in
    America*. University of Chicago Press
Kanter, Rosabeth Moss. 1977. *Men and Women of the Corporation*. New York:
    Basic Books
    1984. "Variations in Managerial Career Structures in High Technology Firms." In
    Paul Osterman (ed.), *Internal Labor Markets*, pp. 109–131. Cambridge, MA:
    MIT Press
Karpick, Lucien. 1978. *Organization and Environment: Theory, Issues and Reality*.
    London: Sage Publications
Kerr, Clark, T. Dunlop, F. Harbison, and C. A. Myers. 1962. *Industrialism and
    Industrial Man*. London: Heinemann. (First published 1960.)
Kindleberger, Charles P. 1976. "Technical Education and the French Enter-
    preneur." In Ed. C. Carter et al. (eds.), *Enterprise and Entrepreneurs in
    Nineteenth and Twentieth Century France*. Baltimore: Johns Hopkins University
    Press
    1978. "Germany's Overtaking of England 1806–1914." In his *Economic Response:
    Comparative Studies in Trade, Finance and Growth*. Cambridge, MA: Harvard
    University Press
Kirchheimer, Otto. 1966. "Private Man and Society." *Political Science Quarterly*, 81:
    1–24
Kocka, Jurgen. 1980. *White Collar Workers in America, 1890–1940*. Beverly Hills,
    CA: Sage Publications
Kolboom, Ingo. 1982. "Patronat et cadres: la contribution patronale à la formation
    du group des cadres (1936–1938)." *Mouvement Social*, 121 (October–
    December), 71–95
Kornblum, William. 1974. *Blue Collar Community*. University of Chicago Press
Kornhauser, William. 1965. *Scientists in Industry*. Berkeley: University of California
    Press
Kraft, Philip. 1977. *Programmers and Managers: The Routinization of Computer
    Programming in the United States*. New York: Springer-Verlag
Laffitte, Pierre. 1973. *Les Ecoles d'ingénieurs en France*. (Notes et Etudes Docu-
    mentaires, no. 4045–4046–4047) Paris: La Documentation française
Lamirand, Georges. 1954. *Le Rôle social de l'ingénieur*. Paris: Plon. (First pub-
    lished 1932.)
Lane, Robert. 1962. *Political Ideology*, Glencoe, IL: The Free Press
    1966. "The Decline of Politics and Ideology in a Knowledgeable Society."
    American Sociological Review, 31: 649–662
Larson, Magali Sarfatti. 1977. *The Rise of Professionalism*. Berkeley: University of
    California
    1980. "Proletarianization and Educated Labor." *Theory and Society*, 9: 1
    (January): 131–175
Lasserre, Henri. 1984. "Systèmes de représentations et idéologies des ingénieurs
    français." *Culture Technique*, 12 (March): 239–245

264   List of references

Layton, Edwin T. 1971. *The Revolt of the Engineers*. Cleveland: Case Western University Press

LeFranc, Georges. 1950. *Les Expériences syndicales en France de 1939 à 1950*. Paris: Editions Montaigne

1967. *Le Mouvement syndical sous la Troisième République*. Paris: Payot

1969. *Le Mouvement syndical de la libération aux événements de mai–juin 1968*. Paris: Payot

Le Guen, Jean-Paul. 1982. "Le Sacrifice: un interview avec Paul Marchelli." *Nouvelle Usine*, 9 September, no. 37

Le Monde de l'Education, no. 105. 1984. An issue featuring several articles under the general heading, "Devenir ingénieur", by Christine Garin and Yves-Marie Labe

Levy-Leboyer, Maurice. 1979. "Le Patronat français, 1912–1973." In Levy-Leboyer (ed.), *Le Patronat de la Seconde Industrialisation*, pp. 137–187. Paris: Les Editions ouvrières

1980. "The Large Corporation in Modern France." In A. Chandler and H. Daems (eds.), *Managerial Hierarchies: Comparative Perspectives on the Rise of the Industrial Enterprise*, pp. 117–170. Cambridge, MA: Harvard University Press

Lipset, Seymour Martin. 1967. "The Changing Class Structure and Contemporary European Politics." In Stephen Graubard (ed.), *A New Europe?*, pp. 337–369. Boston: Beacon Press

Lipset, Seymour Martin, M. A. Trow, and J. F. Coleman. 1956. *Union Democracy: The Internal Politics of the International Typographical Union*. Glencoe, IL: The Free Press

Littler, Craig R. 1981. *Power and Ideology in Work Organization: Britain and Japan*. Milton Keynes, England: Open University Press

Lortie, Dan C. 1975. *Schoolteacher: A Sociological Study*. University of Chicago Press

Loveridge, Ray. 1983. "Sources of Diversity in Internal Labor Markets." *Sociology*, 17: 1 (February): 44–78

Low-Beer, John. 1978. *Protest and Participation: The New Working Class in Italy*. Cambridge University Press

1981. "Cultural Determinism, Technological Determinism, and the Action Approach." *Research in the Sociology of Work*: 1: 403–433. Jai Press, Inc.

Mackenzie, Gavin. 1973. *The Aristocracy of Labor*. Cambridge University Press

1974. "The 'Affluent Worker' study: an evaluation and critique." In F. Parkin (ed.), *The Social Analysis of Class Structure*. London: Tavistock

Maier, Charles. 1970. "Between Taylorism and Technocracy: European Ideologies and the Vision of Industrial Productivity in the 1920s." *Journal of Contemporary History*, 5: 27–61

Mallet, Serge. 1975. *The New Working Class*. Nottingham, England: Spokesman Books. (Published in French, 1963; 5th edn 1969)

Marcson, S. 1969. *The Scientist in American Industry*. New York: Harper and Row

Marcuse, Herbert. 1964. *One Dimensional Man*. Boston: Beacon Press

Martin, Roger. 1984. *Patron de droit divin*. Paris: Gallimard

Maurice, Marc. 1969. "Professionnalisme et syndicalisme." *Sociologie du Travail*, 3: 243–256

1971. "The Evolution of French Cadres and the Collective Action." *International Studies of Management and Organizations* (Spring)

Maurice, Marc and R. Cornu. 1970. "Revendications, orientations syndicales et participation des cadres à la grève." *Sociologie du Travail*, 12 (3): 328–337

Maurice, Marc, F. Sellier, and J.-J. Sylvestre. 1977. *Production de la hiérarchie dans l'entreprise: recherche d'un effet social, Allemagne–France*. Aix-en-Provence, France: Laboratoire d'économie et de sociologie du travail

1984. "The Search for a Societal Effect in the Production of Company Hierarchy: A Comparison of France and Germany." In P. Osterman (ed.), *Internal Labor Markets*, pp. 231–270. Cambridge, MA: MIT Press

1986. *The Social Foundations of Industrial Power: A Comparison of France and Germany*. Cambridge, MA: MIT Press

Meiksins, Peter F. 1982. "Science in the Labor Process: Engineers as Workers." In Charles Derber (ed.), *Professionals as Workers: Mental Labor in Advanced Capitalism*. Boston: G. K. Hall and Co.

Meiksins, Peter F. and James Watson. 1987. "Autonomy and the Engineer: The Degradation of Professional Work?" Paper presented to the SSSP Meeting, Chicago, IL, August 14

Mills, C. Wright. 1951. *White Collar: The American Middle Classes*. New York: Oxford University Press

Monjardet, Dominique. 1972. "Carrière des dirigeants et contrôle de l'entreprise." *Sociologie du Travail*, 14 (2): 131–44

1979. "Les Cadres de l'industrie, les classes sociales et la sociologie." Unpublished manuscript

1980. "Organisation, technologie et marche de l'entreprise industrielle." *Sociologie du Travail*, 1: 76–96

Montagna, Paul. 1977. *Occupations and Society*. New York: Wiley

Mortimer, Jeylan T. and J. London. 1984. "The Varying Linkages of Work and Family." In Patricia Voydanoff (ed.), *Work and Family*, pp. 20–35. Palo Alto, CA: Mayfield Publishing Company

Mouriaux, René. 1984. "Le Syndicalisme des ingénieurs et cadres." *Culture Technique*, 12 (March): 221–228

Moutet, Aimée. 1985. "Ingénieurs et rationalisation en France de la guerre à la crise (1914–1924)." In A. Thépot (ed.), *L'Ingénieur dans la société française*, pp. 71–108. Paris: Les Editions ouvrières

Naville, Pierre. 1963. *Vers l'automatisme social?* Paris: Gallimard

Noble, David. 1977. *America by Design: Science, Technology and the Rise of Corporate Capitalism*. New York: Alfred Knopf

Nouvelle Usine. 1982. "Palmarès 1982: Des écoles d'ingénieur électronique." 27 March, no. 22

O'Hara, Patrick. 1985. "L'Industrie électronique: le savoir et le chronomètre." *Société Française*, 14: 11–24

Osterman, Paul. 1984. *Internal Labor Markets*. Cambridge, MA: MIT Press

Papanek, Hannah. 1973. "Men, Women and Work: Reflections on the Two-person Career." *American Journal of Sociology*, 78: 852–872

Parkin, Frank. 1979. *Marxism and Class Theory: A Bourgeois Critique*. New York: Columbia University Press

Parsons, Talcott. 1949. "The Professions and Social Structure." In his *Essays in Sociological Theory, Pure and Applied*. Glencoe, IL: The Free Press

Pavalko, Ronald M. 1988. *Sociology of Occupations and Professions*. 2nd edn, Itasca, IL: F. E. Peacock Publishers, Inc.

Perrow, Charles. 1979. *Complex Organizations: A Critical Essay*. 2nd edn, Glenview, IL: Scott, Foresman and Co.

Perrucci, Robert and Gerstl, Joel. 1969. *Profession Without Community: Engineers in American Society*. New York: Random House

Petitjean, Gérard. 1986. "Grandes écoles: trop petites . . . " *Le Nouvel Observateur* (31 January–6 February), pp. 52–55

Poulantzas, Nicos. 1978. *Classes in Contemporary Capitalism*. London: Verso

Prandy, Kenneth, A. Stewart and R. Blackburn. 1982. *White Collar Work*. London: Macmillan

Prost, Antoine. 1968. *L'Enseignement en France 1800–1967*. Paris: Armand Colin

Quin, Claude. 1976. *Classes sociales et union du peuple de France*. Paris: Editions sociales

Reid, Donald. 1985. "Industrial Paternalism: Discourse and Practice in Nineteenth-Century French Mining and Metallurgy." *Comparative Studies of Society and History*, vol. 27: no. 4 (October 1985), 579–601

Renner, Karl. 1978. "The Service Class." In T. Bottomore and P. Goode (eds.), *Austro-Marxism*, pp. 249–252. Oxford: Clarendon Press

Reynaud, Jean-Daniel. 1972. "La Nouvelle Classe ouvrière, la technologie et l'histoire." *Revue Française de Science Politique*, 22 (3): 529–542

1975. *Les Syndicats en France*. Paris: Editions du Seuil

Ribeill, Georges. 1984. "Entrepreneurship hier et aujourd'hui: la contribution des ingénieurs." In A. Grelon (ed.), *Culture Technique*, 12 (March): 77–90

Ringer, Fritz K. 1979. *Education and Society in Modern Europe*. Bloomington: Indiana University Press

Ritti, Richard R. 1968. "Work Goals of Scientists and Engineers." *Industrial Relations*, 7: 118–131

1971. *The Engineer in the Industrial Corporation*. New York: Columbia University

Ritzer, George. 1986. *Working: Conflict and Change* (3rd edn). Englewood Cliffs, NJ: Prentice-Hall

Rivard, P., J.-M. Saussois, and P. Tripier. 1982. "L'Espace de qualification des cadres." *Sociologie du Travail*, 4: 417–442

Ross, George. 1978. "Marxism and the New Middle Classes: A French Critique." *Theory and Society*, 5 (2): 163–192

Rothstein, William G. 1969. "Engineers and the Functionalist Model of the Professions." In R. Perrucci and J. Gerstl (eds.), *The Engineers and the Social Structure*, pp. 73–98. New York: Wiley

Rubery, Jill. 1987. "Employers' Strategies and the Labour Force." Unpublished

Rubin, Lillian B. 1976. *Worlds of Pain: Life in the Working Class Family*. New York: Basic Books

Sabel, Charles. 1982. *Work and Politics: The Division of Labor in Industry*. New York: Cambridge University Press

Saglio, Jean. 1984. "Les ingénieurs sont-ils des patrons comme les autres?" *Culture et Technique*, 12 (March): 93–101, 349

Sainsaulieu, Renaud. 1972. *Les Relations du travail a l'usine*. Paris: Les Editions d'organisation

Savage, Dean. 1979. *Founders, Heirs and Managers: French Industrial Leadership in Transition*. Beverly Hills, CA: Sage Publications

Scott, W. Richard. 1981. *Organizations: Rational, Natural and Open Systems*. Englewood Cliffs, NJ: Prentice-Hall

Sewell, William H. Jr. 1980. *Work and Revolution in France*. New York: Cambridge University Press

Shinn, Terry. 1980a. "From 'Corps' to 'Profession': The Emergence and Definition of Industrial Engineering in Modern France." In Robert Fox and George Weisz (eds.), *The Organization of Science and Technology in France 1808–1914*, pp. 183–208. Maison des Sciences de l'Homme and Cambridge University Press
    1980b. *Savoir scientifique et pouvoir sociale*. Paris: Presses de la Fondation Nationale de Science Politique

Smith, Chris. 1987. *Technical Workers: Class, Labour and Trade Unionism*. London: Macmillan

Sofer, Cyril. 1970. *Men in Mid Career: A Study of British Managers and Technical Specialists*. Cambridge University Press

Sorge, Arndt. 1979. "Engineers in Management: A Study of the British, German and French Traditions." *Journal of General Management*, 5: 46–57

Sorge, Arndt and M. Warner. 1986. *Comparative Factory Organization: An Anglo-German Comparison of Management and Manpower in Manufacturing*. Berlin: Gower

Spenner, Kenneth. 1983. "Temporal Changes in the Skill Level of Work." *American Sociological Review*, 48: 6 (December), 824–837

Spilerman, Seymour. 1977. "Careers, Labor Market Structure, and Socioeconomic Achievement." *American Journal of Sociology*, 83: 551–593

Stark, David. 1980. "Class Struggles and the Transformation of the Labor Process: A Relational Approach." *Theory and Society*, 9: 89–130
    1986. "Rethinking Internal Labor Markets: New Insights from a Comparative Perspective." *American Sociological Review*, 51: 4 (August): 492–504

Stearns, Peter N. 1978. *Paths to Authority: The Middle Class and the Industrial Labor Force, 1820–1848*. Urbana: University of Illinois

Stewart, A., K. Prandy, and R. M. Blackburn. 1980. *Social Stratification and Occupations*. London: Macmillan

Stinchcombe, Arthur L. 1965. "Social Structure and Organization." In J. March (ed.), *Handbook of Organizations*, pp. 142–169. Chicago: Rand McNally
    1978. *Theoretical Methods in Social History*. New York: Academic Press

Suleiman, Ezra N. 1978. *Elites in France*. Princeton University Press

TELECO. 1976. *Annual Report*

Thépot, André. 1979. "Les Ingénieurs du corps des mines, le patronat, et la seconde industrialisation." In Levy-Leboyer (ed.), *Le Patronat de la seconde industrialisation*, pp. 237–246. Paris: Les Editions ouvrières

1985. *L'Ingénieur dans la société française*. Paris: Les Editions ouvrières

Tilly, Charles. 1978. *From Mobilization to Revolution*. Reading, MA: Addison-Wesley

Tocqueville, Alexis de. 1945. *Democracy in America*, Bradley Edition, vol. 2. New York: Vintage Books

Torstendahl, Rolf. 1982. "Engineers in Industry, 1850–1910: Professional Men and New Bureaucrats, A Comparative Approach." In C. G. Bernhard et al. (eds.), *Science, Technology and Society in the Time of Alfred Nobel*. Oxford: Pergamon Press

Touraine, Alain. 1971. *The Post-Industrial Society*. New York: Random House

Trilling, Leon. 1979. "Technological Elites in France and the United States." *Minerva*, 17 (2): 225–243

Udy, Stanley. 1981. "The Configuration of Occupational Structure." In Hubert Blalock, Jr. (ed.), *Sociological Theory and Research*. Glencoe, IL: The Free Press

UIMM (Union des industries métallurgiques et minières). 1977. *Ingénieurs et cadres des industries des métaux: situation au 1-1-1976 prévision de besoins*

Veblen, Thorstein. 1965. *The Engineers and the Price System*. New York: Viking. (First published 1921)

Vincent, Gérard. 1977. *Les Français 1945–1975*. Paris: Masson

Vollmer, Harold and Donald Mills. 1966. *Professionalism*. Englewood Cliffs, NJ: Prentice-Hall, Inc.

Weiss, John. 1982. *The Making of Technological Man: The Social Origins of French Engineering*. Cambridge, MA: MIT Press

Whalley, Peter. 1986. *The Social Production of Technical Work: the Case of British Engineers*. London: Macmillan; Albany: State University Press

Wilensky, Harold. 1960. "Work, Careers, and Social Integration." *International Social Science Journal*, 12: 543–560

1964. "The Professionalization of Everyone?" *American Journal of Sociology*, 70: 137–158

Willener, Alfred et al. 1969. *Les Cadres en mouvement*. Paris: Editions de l'Epi

Windolf, Paul. 1983. "L'Expansion de l'enseignement et la surqualification sur le marché du travail." *Archives Européenes de Sociologie*: 101–143

Wood, Stephen (ed.). 1982. *The Degradation of Work?* London: Hutchinson

Woodward, Joan. 1958. *Management and Technology*. London: Her Majesty's Stationery Office

1965. *Industrial Organization: Theory and Practice*. New York: Oxford University Press

Wright, Erik Olin. 1979. *Class Crisis and the State*. London: Verso

1980. "Class and Occupation." *Theory and Society*, 9: 177–214

1985. *Classes*. London: Verso

Wylie, Laurence. 1964. *Village in the Vaucluse*. New York: Harper and Row. (First published in 1957.)

Zeldin, Theodore. 1979. *France 1848–1945 Ambition and Love*. Oxford University Press

1982. *The French*. New York: Pantheon Books

Zussman, Robert. 1985. *Mechanics of the Middle Class: Work and Politics among American Engineers*. Berkeley: University of California Press

Zysman, John. 1977. *Political Strategies for Industrial Order: State Market and Industry in France*. Berkeley: University of California Press

Index of names

Subject index

aeronautical industry, 2, 14, 65, 87, 136
age
  career lines and, 16, 141
  political preference and, 215–216
  salaries and, 148
alumni associations, 155, 189
  labor market control and, 13, 194
  membership in, 187
American engineers
  career mobility for, 4–5, 141–143, 155
  collective representation for, 195
  empirical studies of, 1, 2, 13, 14, 15, 19,
    131, 228
  politics of, 197, 226–227
  residential patterns of, 201, 206
  technical education for, 57, 58
  work situation for, 71–72, 79, 106
apprenticeship system, 72, 117, 154
Arts et Métiers, Ecole Nationale Supérieur
  des, 55–56, 73, 136
Association pour l'emploi des cadres
  (APEC), 155
associations, *see* professional associations
authority, 100–133
  challenges from engineers to, 127, 132–133
  communications flow and, 105
  engineers as commanders model and,
    108–109, 131–132
  indirect supervision and, 106–113
  ingénieurs and connection to, 124–125
  operation and perception of, 106–113
  participation by engineers and, 117
  perceptions and criticisms of, 126–127,
    131–132
  professional class and, 10–11
  professionalization and, 4
  social class and, 223

technical education and, 61
technical workers and challenges to, 1
uses of, 121–127
autodidactes, see ingénieurs autodidactes
autogestion
  attitudes toward, 120–121
  new working class theory and, 9
  trade unions and, 172, 177
automobile industry, 65, 73
autonomy, 117, 217
  deskilling and, 13, 14
  experience and bases of, 100–106
  freedom from direct supervision and,
    101–103
  freedom of movement and, 104
  limits of, 110–111
  new working class theory on, 8, 9
  organizational structure and, 103–106
  product engineering and, 231
  professionalization and, 4, 5
  proletarianization and, 6
  TELECO organisational structure with,
    87

Belgium, 149
Brevet de Technicien Supérieur (BTS), 58,
  139, 153
British engineers
  autonomy of, 106, 112
  career advancement for, 141–143, 155
  collective representation for, 195
  empirical studies of, 1, 2, 13, 14, 15, 131,
    228
bureaucracy, 132
  Pierre's interview on, 43–44
  professionalization and, 4
  TELECO and, 26

273